HOT MESS

MOTHERING THROUGH A CODE RED CLIMATE EMERGENCY

SARAH MARIE WIEBE

HOT MESS

MOTHERING THROUGH A CODE RED CLIMATE EMERGENCY

SARAH MARIE WIEBE

FOREWORD BY
Rachel yacaaʔał George

ILLUSTRATIONS BY
Indi Maverick

Copyright 2024 © Sarah Marie Wiebe

All rights reserved. No part of this book may be reproduced or transmitted in any form by any means without permission in writing from the publisher, except by a reviewer, who may quote brief passages in a review.

Development editor: Tanya Andrusieczko
Copyediting: Amber Riaz
Cover design: Evan Marnoch
Interior illustrations: Indi Maverick
Text design: Lauren Jeanneau
Printed and bound in the UK

Published by Fernwood Publishing
Halifax and Winnipeg
2970 Oxford Street, Halifax, Nova Scotia, B3L 2W4
www.fernwoodpublishing.ca

Fernwood Publishing Company Limited gratefully acknowledges the financial support of the Government of Canada through the Canada Book Fund and the Canada Council for the Arts. We acknowledge the Province of Manitoba for support through the Manitoba Publishers Marketing Assistance Program and the Book Publishing Tax Credit. We acknowledge the Nova Scotia Department of Communities, Culture and Heritage for support through the Publishers Assistance Fund.

Library and Archives Canada Cataloguing in Publication
Title: Hot mess : mothering through a code red climate emergency / Sarah Marie Wiebe.
Names: Wiebe, Sarah Marie, author
Description: Includes bibliographical references and index.
Identifiers: Canadiana (print) 20240418964 | Canadiana (ebook) 20240423194 | ISBN 9781773635668 (softcover) | ISBN 9781773637129 (EPUB) | ISBN 9781773637112 (PDF)
Subjects: LCSH: Wiebe, Sarah Marie. | LCSH: Motherhood. | LCSH: Parenting—Political aspects. | LCSH: Ecofeminism. | LCSH: Climatic changes—Physiological effect. | LCSH: Climatic extremes.
Classification: LCC HQ759 .W54 2024 | DDC 306.874/3—dc23

Contents

FOREWORD
What It Means to (Re)Centre Care 1

PREFACE
A State of Emergency as an Embodied Event..6

CODE RED
Feminist Motherhood in a World on Fire... 27

CODE ORANGE
Cultivating Community through Disasters...39

CODE PINK
A Splash and Dash Caesarean 53

CODE BLUE
Living through Multiple Crises, Climate Anxiety, and Mental Health 67

CODE GREEN
Circular Economies of Care..85

CODE BLACK
Systemic Threats, Revealing Violences Slow and Spectacular 105

CODE GREY
A Cautionary Tale of Renewable Extraction 121

PRISMATIC REFLECTIONS
Cultivating Care and Community through Multifaceted Crises............. 134

Acknowledgements.. 148

Endnotes 151

Index.. 177

For Forest

FOREWORD

What It Means to (Re)Centre Care

WHEN THE HEAT DOME COVERED WESTERN CANADA in the summer of 2021, I was nearing the end of the second trimester of my pregnancy. I had already been experiencing some discomfort as my daughter had begun digging her tiny feet beneath my ribs, and the threat of the heat wave intensified these feelings in a condition — as Sarah Wiebe similarly describes — of "sweaty, sticky disarray." During this time, my husband and I were living in Amiskwaciwâskahikan (Edmonton) in Treaty Six territory. As temperatures climbed in the city to the high 30s and low 40s, I exchanged screenshots of daily weather app updates and made frequent phone calls to friends and family in British Columbia's Lower Mainland and Victoria as I prayed for their safety amid the unrelenting heat. The weight of anxiety bore down on me as I feared for family struggling to stay cool. We could only watch what was happening from across the Rocky Mountains while we battled our own high temperatures. We were fortunate enough to have air conditioning in our home, so we opened our space to our friends in the city. My dear friend, in the final weeks of her own pregnancy, napped in our basement to beat the heat that had overtaken her home. As Wiebe echoes throughout *Hot Mess,* these moments of crisis place care at the centre, emphasizing the ways we come through for one another in the direst of times. At the same time, these moments also remind us of how practices of care are vital to the survival of the multitude of life at any point—not just in crisis—thus emphasizing the necessity of centring our teachings as Indigenous peoples.

My daughter was born on the prairies while we lived as visitors for three and a half years in Amiskwaciwâskahikan. Despite our landlocked environment while in Treaty Six, it was the sound of the ocean that her spirit craved, of all the sounds that could have possibly soothed her. It was the sound of waves that calmed her and lulled her to a peaceful sleep, which didn't always come easily when she was an infant. She came into the world moving through water, transitioning from the amniotic fluid that surrounded her to the warm pool and then into my arms on a crisp morning. On her fourth day, we followed the teachings of our elders and welcomed her to the world with a cedar bath. Cedar is a sacred medicine, and we have been taught the power and healing of water, so bringing these two together brought further sacredness into her first moments, locating her in the world with our medicines. Several months later, when she first dipped her toes in the ocean waters, she looked both at peace and at home. A calmness settled over her, a recognition of belonging. These seemingly small moments remind me that she carries the coast in her, as do we all as coastal Indigenous peoples. It is a reminder of the ways we become tethered to our homelands and waters, bound by the life force of place.

I am tasked with teaching her our ways as nuučaańuł people; to support and guide her in living her life as quuʔas. Our homelands and waters will anchor her as they do me. They will call her home — gentle waves kissing the shoreline, waters conveying messages from timeless guides. She will foster a relationship with place filled with the responsibilities called for by our teachings. Central to this are the practices of care that are integral to our teachings of and around hišukʔiš c̓awaak — everything is one. This philosophy calls on us as Indigenous peoples to remain active participants in a reciprocal relationship with all of creation, asking us to reflect on the spirit in everything and the ways in which we are connected in webs of kinship.

Centring hišukʔiš c̓awaak demands a reflection on our accountability and the ways we carry out our responsibilities in the relationships we hold interpersonally and with our more-than-human relations. Our responsibilities require deep care for our relations practised daily. We locate ourselves as but one life in a web where each life depends on the survival, health, and well-being of the other. These modes of care resist the ways the settler-colonial state has attempted to divide us from one another in emphasizing individualism and extractive relations. When we reflect on our engagement with the world being one of ongoing and renewable relationships of reciprocity, we shift the focus away from destructive discourses embedded in colonization and extraction, and we embody the care necessary to continue envisioning the future. When we move through life holding these teachings as central, we not only extend care to our relations, but we fundamentally embody sustainable relations that honour the life in all of creation. These are the ways of relating to the multitude of life that I will instill in my own daughter, but that also hold relevance beyond this intimate relationship and have the capacity to fundamentally reshape our future globally.

The role of care remains a focal point throughout *Hot Mess,* as Wiebe emphasizes the crises of care exacerbated by states of emergency in a climate crisis. In doing so, she draws attention to the ways emergency patterns highlight the dividing lines between life that can and cannot be grieved and illuminates precisely what is at stake if society continues to fail to grapple with extractive relationships. Through a moving personal recounting of becoming a mother in a Code Red climate emergency, Wiebe explores what it would look like to push back against Western extractive and transactional approaches to our living environments and, instead, to embody the care called for in many Indigenous teachings about our relationships to the world. Echoing the calls by Indigenous scholars, activists, and community members for generations, Wiebe posits this as a shift that would fundamentally alter society's engagement with the world for the better. This collection of personal reflections serves as a poignant call to action to (re)centre care. I bracket off *re-* in *(re)centring care* to draw attention to the fact that care has been an ongoing integral practice across many Indigenous communities since time immemorial. Indeed, as noted earlier, care is foundational in many of the teachings in our respective nations. *Recentring* (without the parentheses) can be taken to suggest a point in time when care was not pivotal. This usage is not my intent. Instead, I offer both options — recentring and centring — as a way of drawing attention to the various ongoing and revitalized practices of care necessary to our survival.

While life during a climate emergency is layered and complicated, taking the "mutually beneficial relations between a parent and infant as an organizing symbol of our political relations" opens space for the creative envisioning of futures that are generative, healthy, and sustaining.

— Rachel yacaaʔał George
Professor, University of Victoria

PREFACE

A State of Emergency as an Embodied Event

THE YEAR I BECAME A MOTHER — 2021 — meteorologists warned that the heat dome trapping us in my home province of British Columbia would be deadly. For three days in a row, temperatures broke records by reaching 20 to 25 degrees above seasonal average in cities across the province, peaking at 49.6°C in Lytton — high enough to provoke nearby wildfires to burn the village to the ground.[1] Years later, local residents continue to live with the aftermath, trying to rebuild their communities. Reports reveal how heat records continue to be broken worldwide and that the amplification of wildfires will lead to even more smoke.[2]

The heat dome was the deadliest weather event in Canadian history, with the BC Coroner's report documenting 619 heat-related deaths.[3] Most who died were elders with compromised pre-existing health conditions, living in isolation. More of the decedents lived in socially and materially deprived neighbourhoods, in comparison with the rest of British Columbia's population, with inadequate cooling infrastructure.[4] This event signalled that the climate emergency is a matter of life and death and that we are not prepared. Health authorities across the province now note that climate change events, ranging from wildfire smoke to droughts, heat, storms, and flooding, correspond with deadly health risks and signal an "existential threat."[5] The heat dome prompts critical reflection about the gradients of life between these extremes, about how people came together to care for one another and survive this highly consequential event for community and planetary health.

Such high heat can lead to loss of consciousness, organ failure, and death. It also exposes the hard truths that people with less greenery and those living below the poverty line or living alone are more likely to die from extreme heat. People with complex mental health concerns such as schizophrenia are even more vulnerable.[6] Waiting times to get through to 911 and then for an ambulance to arrive were contributing factors to increased vulnerabilities. 911 calls doubled during the peak of the heat dome, between June 25 and July 1, 2021.[7] In six instances, callers were informed that no ambulances were available. British Columbians learned some hard truths that summer about the pressing reality of the climate emergency, notably that we need to be better prepared and to check on one another, especially those considered vulnerable and those living alone.

With my six-week-old baby and partner in tow, I tried to keep my cool during the exceptionally hot first summer of my baby's life. We dipped our feet into the ocean, ran an ice-water bath, hung out in a kiddie pool with our neighbours, and knotted frozen bandanas around our heads, the sweat impervious to our strategies. In those early, fragile days of motherhood, I was still struggling to find my groove as a nursing parent. Becoming a mother is already an unravelling of one's sense of composure and identity; added to this was the hot mess of a dramatic, extreme weather event. This unnerving, chaotic feeling of becoming a sticky hot mess shaped the panicky early days of motherhood so much that it felt like a metaphor for the dramatically changing world my son was born into.

While riding out the heat wave, after the exceptionally hot June of 2021, I underestimated the amount of water I needed to keep not only my body, but my son's body hydrated. Just a few months after a traumatic emergency C-section birth, I again found myself admitted to a hospital emergency ward, awaiting care for unrelenting migraines that began during the heat dome.[8] The intense external weather pressure system mirrored and tested the limits of my internal pressure system.

After a bout of nausea that prevented me from nursing our son early one morning, and calls to every urgent care clinic in a 25-kilometre radius, my partner insisted we take a trip to the hospital. I joined the queue for triage that extended through the sliding doors and spilled onto the parking lot, and sat in a jam-packed waiting room for two hours before I saw the emergency doctor. The nurses could not find my veins to give me necessary fluid and medications to abort — not prevent — the incessant migraine episodes. "You have to go to your family doctor for that," was the emergency doctor's response when I pressed her about preventive medications. If only we had a family doctor, I thought to myself, on the brink of tears. The shortage of physicians is another crisis consuming Victoria, where we live.

That day in the emergency ward, I was away from my son for the first time since his birth from dawn till dusk. I learned some tough lessons about the nursing maternal body, most notably how much water one must take in to sustain another life. I became aware of just how intimately the current climate emergency can be felt in the core of our beings. I felt the crisis of care in my body, through my skin, twinned with the visceral sensations of what it feels like when we forget to care about our warming planet.

From my vantage point as a critical ecofeminist and new mother, I am concerned with and focused on how personal sensations are political, how states of emergency make the visceral realities of democratic politics affective, fleshed out in our bodies, on our skin, felt through our kin relationships. Political theorist Tim Luke, in *Anthropocene Alerts,* points out that global emergencies such as a pandemic or the climate emergency call for "an investigation of the relations between democracy, public engagement and policy."[9] I share his concerns. There is a dire need to look more closely at the democratic conditions of and the political environments *leading up to* official State of Emergency declarations — presenting a moment to interrogate and intervene in conversations about the administrative rationalities of emergency politics before it's too late. Most often, State of Emergency declarations address exceptional moments that reinforce executive forms of hierarchical power relations with limited avenues for public engagement and input.

Emergency events and declarations are as much political as they are personal. My emergency Caesarean section took less than fifteen minutes to perform — a terrifying event in the moment, yet a common one. Approximately one in three women birth their babies through an elective or emergency C-section in the United States, and about one in five globally.[10] I share my experiences about navigating parenthood in an intensely changing climate in an effort to contribute to personal and political conversations about the need for more caring relations between and among humans and the planet.[11] Weaving lived experiences, research, and political events, this book reflects on how we can care for each other during crises, especially when states of emergency are declared by political representatives.

Emergency events prompt hospitals and health authorities to declare certain codes to indicate unique sources of trauma. As an example, Code Orange is used to respond to a disaster. During multiple visits to the Royal Jubilee Hospital in Victoria near our home, I noted the Code Orange preparedness signs and emergency kits stashed behind the reception area. When the province of British Columbia declared a State of Emergency due to the rapidly spreading wildfires in the Okanagan in the summer of 2023, the Kelowna General Hospital and Interior Health authority declared a rare Code Orange to care for people fleeing wildfires. This declaration drew sharply into focus just how poignant, personal, and telling such declarations can be for human and planetary health.[12] Such declarations are meant as a rarely used response to disasters or mass casualties.

A central question of this book asks: How do we care for each other, live well on this planet together, and nurture the conditions for life to flourish when exceptional emergency events become the norm? I argue that it requires thinking about life beyond individualism. We are all better off when we care for each other, and this is especially true during emergency events. Moreover, I build upon critical ecofeminist, environmental justice, and Indigenous decolonial thought to highlight how living well for community and planetary well-being requires a radical shift away from extractive capitalism — continuously taking from the earth for profit — and instead, calls for refocusing on reciprocal relationships with the vital environments that sustain life.

I feel the planetary crisis acutely as I attempt to process an inferno of climate grief: during another hot summer in 2023, the province of British Columbia declared a State of Emergency[13] due to unprecedented wildfire conditions, while entire cities like Yellowknife, Northwest Territories (Canada), and Maui, Hawai'i, burn — drawing sharply into focus why global leaders and scientists call the climate crisis a "Code Red emergency" — where the consequences of global heating become more extreme, manifest in events such as the spread of wildfire.[14] Since I started writing about the politics of State of Emergency declarations more than a decade ago during the emergence of the 2012 Idle No More movement, I began to pay attention to emergencies near and far:[15] State of Emergency declarations made by public officials, including substandard housing conditions and mental health crises in northern communities like Attawapiskat; Hawai'i Governor

Ige's declaration against the protesters at Mauna Kea; former US president Donald Trump's efforts to build a wall along the United States–Mexico border; the public health declarations around the world due to the COVID-19 pandemic; and the internationally recognized climate emergency. Emergency declarations seem to be the new normal.[16] As critical policy scholar Michael Orsini writes, even grief due to the "avalanche of death" starts to become ordinary.[17] Emergency conditions are matters of life and death. Our human bodies and bodies of water, along with our environments, are all at the frontlines.

A State of Emergency declaration is an extraordinary bureaucratic moment, a sovereign or authoritative decision, with consequential temporary rules and proceedings — often understood as a request for help from other orders of government. It can be declared by any governing body — be it municipal, First Nation, provincial, federal, or international. A State of Emergency can be understood as a policy assemblage: a bureaucratic entanglement of institutions, discourses, practices[18] that are simultaneously symbolic and highly consequential for governance. The normal rules of proceedings can be suspended, and status quo democratic practices and norms can be temporarily suspended.[19] This tension came to light in Hawaiʻi when elected officials sought to declare a housing State of Emergency on July 17, 2023, leapfrogging over normal permitting procedures, and expediting certain forms of development without comprehensive civic consultation or input.[20]

State of Emergency declarations often cast a light on some of the hidden, systemic inequalities that underpin the asymmetrical power relations of liberal democracies themselves. In some instances, such as has been the experience of Attawapiskat leadership in Northern Ontario — a community affected by housing shortages, a youth mental health crisis, and water contamination — State of Emergency declarations reveal affected communities' underlying concerns with environmental injustice and significant gaps in Canada's respect for Indigenous self-determination, rights, and treaty relations. These policy instruments have the power to challenge status quo governance and rules of proceeding, to serve as a call to action for democratic life otherwise, and to signal a pivotal moment to collectively reimagining healthier futures.

Such declarations raise questions about what it means to live through and survive a State of Emergency. These declarations affect personal circumstances specifically and democratic life more broadly — who has (and does not have) the power to make these declarations? State of Emergency declarations press us to flesh out the parameters for a healthier body politic.[21] Living through State of Emergency conditions prompts reflection about the consequences of emergencies for equity and how humans relate to each other and to broader ecologies.

Emergencies draw attention to unprecedented, unpredictable events, or so one might assume. It also becomes apparent how the exceptional masks what is routine. As emergencies become normalized, so too do the politics of resource mobilization, where public officials must decide swiftly how to respond to and manage these events. These are not neutral decisions. They are laden with power, significance, and meaning. When emergency patterns emerge, so too do the dividing lines between grievable and nongrievable forms of life — more specifically about whose lives matter most when it comes to protection in a political community.[22] Not everyone experiences a State of Emergency in the same way.

Emergency declarations, events, and experiences hold together complicated emotions. They trouble dualistic ways of thinking. By their very definition, emergencies are vital matters of life and death. At the same time, emergency declarations and events mean much more to those whose lives are directly affected by these vital poles of life or death scenarios. An emergency is generally a response to an unanticipated event. An emergency may signal panic or fear. Simultaneously, emergencies may signal a sense of possibility, for a response and safety. Acute emergencies ignite swift action. But how do we contend with slow emergencies, when the causes are layered and systemic, and thus pose a challenge to diagnose?

The experiences of the day I left my baby behind to be treated in the emergency room shaped my thinking about the pace and feelings generated by emergencies and their variances, be they fast or slow. I had an issue, and later that day it was treated. At the same time, my issue was entwined with the climate emergency. This provoked a deeper visceral concern about how to treat systemic, chronic, long-term slow emergencies. Bodies — human, like the nursing parent — are beacons, offering up important lessons about the root causes of emergency events and the need to reimagine conditions otherwise.

Emergencies are both lived experiences and metaphorical representations of contemporary body politics. When former

Chief of Attawapiskat, Theresa Spence, began her ceremonial fast in 2012, in response to a housing crisis in her community that led to a State of Emergency declaration, her body became a signal for the complicated, layered, multidimensional emotions related to colonial neglect and the breakdown of healthy treaty relationships. While the event catalyzed widespread social mobilization through the Idle No More movement, it also unmasked slow-moving violence, a slow emergency of pain, trauma, and neglect of Indigenous lives in her community and beyond. When both Prime Minister Stephen Harper and Governor-General David Johnston refused to meet with her, they also refused to respond to the emergency. As "heads of state" choose whether to respond to the crisis atmospheres or political emergencies that we experience, the palliative politics they enact deal with some of the immediate pain for some people, but they do not address systemic causes.[23] Emergencies command attention, forcing the triaging of political life itself.

The hospital emergency management system lends an important perspective to our current climate emergency. Hospitals and health authorities have a unique colour-coded messaging system to alert staff to crises. Each colour has a specific warning resonance in the context of a hospital setting with broader implications for human relations with the planet. This book takes on the colour codes to explore the climate emergency. Each code evokes a mood, feeling, or sentiment in relation to a particular emergency condition. The codes have broader interpretive, metaphorical resonance for personal, political, and planetary life. Emergency code sensations are deeply affective; they are as profoundly personal, felt, and visceral as they are power-laden and political. They serve as a literal and metaphorical colour-coded alert system, a signal to address the climate emergency through lived experience.

In 2021, the year my son was born, the United Nations Secretary-General along with scientists and global health officials called climate change a "code red for humanity"; this code sets the scene for the first vignette and title of the book.[24] Code Red, in the hospital system, is often used for the threat of fire. Climate scientists are now calling attention to a global wildfire crisis, along with related extreme weather events and a rapidly warming world.[25] *Hot Mess* discusses the blurry state of becoming a mother in this warming context as both a lived experience of new motherhood and living within our current climate emergency.

The second vignette, "Code Orange," elaborates the relationships between fast and slow disasters. I discuss my research relationships based in Canada's Chemical Valley, a site of slow-moving disasters in a pollution hotspot. This approach is informed by Rob Nixon's work on slow disasters that are types of disasters that often seem so subtle they are unrecognized.[26]

Following this discussion, "Code Pink" elaborates my own experience through an obstetrical emergency and presents details of my personal birth story. Code Pink often refers to a pediatric emergency and/or obstetrical emergency. I discuss the concept of Nuu-chah-nulth teaching yaʔakmis as explained by Indigenous governance scholar and seascapes research collaborator Dawn Smith — and the twin sensations of love and pain.[27] Through these experiences, we learn more about what matters most to us. This concept helps me think about matters of life and death, and to process the twin traumas of a challenging birth and the responsibilities that come with raising a son in this dramatically changing climate.

The next vignette, "Code Blue" discusses climate anxiety, postpartum mental health and encounters with hierarchical, instrumentalist, transactional managerial blue bureaucracies to elaborate upon complex emotions between love of the planet and a painful fear about the future. In the hospital system, a Code Blue often refers to a cardiac arrest and implies a serious life or death threat — I also think of it like a broken heart — the result of emotional or physical trauma, which may result from a sudden illness, the loss of a loved one, a serious accident or a natural disaster.[28] Based on personal and policy examples, I suggest that it can be applied to the pain of processing climate disaster and as a portal to consider waterways in more relational terms, as abundant forms of sea life. As I explain, turning to the ocean, to waterways, these bodies of water can offer human bodies some reprieve. Along with processing anxiety, heartache, pain, and grief, "blue" in this chapter reorients how we think about our bodies, environments, and territories in relation to water. Rigid, hierarchical, instrumentalist forms of blue bureaucracies highlight policy and governance challenges, which, in turn, contend with more relational approaches to waterways and abundant ocean governance, what some of my colleagues and I refer to as "seascapes."[29]

In our capitalist system, which conceives of environments in transactional terms, viewing natural resources as sources of profit, we are encouraged to make individualistic choices about prosperity with little regard for our impact on planetary life. Individual responsibility for sustainable consumer choices branded "green" surround us, but rarely do these narratives unearth the systemic injustices perpetrated through the exploitation of our environments. In the vignette that follows, I aim to abandon or evacuate from superficial sustainability rhetoric, or "greenwashing," and articulate instead the vital importance of finding alternatives to our current status quo corporate, profit driven capitalist economy. Code Green is a hospital code often used to discuss evacuations. Evacuating, or moving beyond extractive capitalism to centre care and economic alternatives to the status quo are discussed through a focus on circular and Indigenous economies that offer local teachings and insights for global economic systems of production and regeneration to centre care over profit.

While there are creative and imaginative ways of thinking about human/more-than-human ways of relating to environments beyond extractivism, this book also draws attention to ongoing forms of violence, including environmental violence and racism that persists in Canada and around the world. "Code Black" draws these violent power relations into focus. As Ingrid Waldron elaborates in *There's Something in the Water*, race is implicated in environmental policymaking — through action and inaction.[30] "Code Orange," as discussed, sheds light on this with a discussion of persistent pollution in Canada's Chemical Valley and the environmental health impacts on the Aamjiwnaang First Nation. As I write on International Women's Day 2024, I am simultaneously preparing my witness testimony for the Senate of Canada in support of Bill C-226: *An Act Respecting the Development of a National Strategy to Assess, Prevent and Address Environmental Racism and to Advance Environmental Justice.* This bill is largely inspired by Waldron's scholarship and would contribute to environmental justice for affected communities across the country.

Environmental racism must be dismantled so that environmental justice can flourish. As explained in the Environmental Noxiousness, Racial Inequities and Community Health (ENRICH) project, led by Waldron, environmental racism includes distributive and procedural injustices, including the inequitable siting of industrial polluters and other environmentally hazardous projects as well as institutional mechanisms and policies that perpetuate inequitable distribution of these activities, which most directly affects racialized communities.[31] In her book, and as elaborated on in "Code Black," Waldron articulates four pillars of critical environmental justice and the Black Lives Matter movement to highlight the vital points of intervention on state-sanctioned forms of structural violence as well as identifies strategies for solidarity and radical transformation of the inequitable status quo. Violence is enacted through institutions, policies, discourses, knowledge systems, and practices with visceral consequences for racialized communities. This vignette concludes with some words of wisdom by bell hooks to connect themes of justice, love, and solidarity with pathways for more caring communities and worlds.

The final vignette blurs the singular code approach and discusses the nuances of colour coding schemes. "Code Grey" — a signal for systems failure — is about current struggles over essential minerals, such as lithium and nickel, required for the electric vehicle revolution. It is an environmental justice conflict poised to define the next generation, hailed as a green strategy for the future, while replicating deadly modes of extraction from lands and waters and marginalized workers. As the provincial government of British Columbia and other governments around the world push for the extraction of critical minerals, taxpayers generally, and local communities in particular, will pay the price for clean-up costs.[32] The power-laden governing systems and development of essential infrastructure for what's purported as clean, green technology both in the electric vehicle manufacturing sector and renewable energy (solar, wind, tidal, geothermal, etc.) come under scrutiny. This critical stance is a cautionary tale for these energy sectors and one that simultaneously seeks to refocus on regenerative human/more-than-human relations rather than extraction.

The phrase "hot mess" is a metaphor for the layered and blurry state of being a new mother and navigating disastrous conditions of contemporary, urgent climate emergency politics. From the Code Pink emergency of birth to the Code Red facing the planet, to Code Orange climate disasters manifesting in both slow and spectacular tempos, "hot mess" highlights that we are living through a vital moment forcing us to rethink our relationships with the planet. In doing this critical and imaginative work, *Hot Mess* responds to the climate emergency by underscoring a caring approach to each other and to the environment while refuting extractive relations. Following Eliane Brum, I begin with the premise that Western, Eurocentric ideals of mastery over nature — typical of modern ways of knowing — have severe consequences for planetary health and the health of human bodies.[33] Capitalist values associated with the dominance of human activity over the natural world has led to the extraction of resources, an overreliance on fossil fuels, and ultimately, the climate emergency that we find ourselves living through. With a deeper diagnosis of the climate emergency's political context, this book simultaneously seeks to challenge the individual blame model of health care, which so often seeps into motherhood shame and blame discourses. Instead, *Hot Mess* calls attention to the layers, textures, and nuances of emergencies and connections we encounter while contending with extractivism, seeking to centre and uphold communities of care.

Care is a radical intervention against the politics of extractive capitalism. In *The End of This World: Climate Justice in So-Called Canada,* contributors argue for the need to unite "to build a caring economy for all."[34] A caring economy requires putting limits on the capitalist economy. This means building a "regenerative economy rooted in care that allows opportunities for meaningful work for all and affords everyone access to the land, water, air, food, education, shelter and community they need to live a good life."[35] *Hot Mess* also calls for placing care at the centre of our economic and political lives.

I aim to open up conversations about issues, about how care is core to living well through and beyond the climate emergency, while raising awareness about the all-too-often silenced intimate matters related to how our bodies respond to emergency interventions: informed consent, emergency C-sections, reproductive mental health, and climate emergency events. By emphasizing care, community, and concern for future generations, I am cautiously optimistic about alternative ecological and decolonial futures, and how we might collectively envision and enact better relations among each other and between humans and the planet, including our atmospheres, lands, waters, animals, and plants, and our living environments.

Emergency events weigh heavily in my heart and on my mind, and they are reinvigorating conversations between critical ecofeminists and ecosocial theorists.[36] Ecofeminism is a body of scholarship that motivates the personal and academic lens that I bring to bear on these lived experiences of layered crises and our current climate emergency. The critical ecofeminist vantage point that I present for this care-centred focus is both a theoretical approach and a frontline activist movement, centred on the politics of community and solidarity for healthy relationships with each other and more-than-human planetary life now and for future generations. Inspired by the concept of felt theory developed by Dian Million, *Hot Mess* draws into focus how situated bodies of knowledge and personal life-experiences are profoundly political.[37] Million's writings remind us that creating space for marginalized voices and experiential knowledges can serve as a radical intervention on patriarchal, capitalist worldviews and perspectives. With this ethic of solidarity, I share my stories alongside the voices of many others to cast light on the viability of other more caring worlds — to show how life beyond extraction can be possible.

Individualistic, neoliberal, capitalist, profit driven ideologies drive our current climate emergency. Moving us out of these dire conditions requires creativity and imagination. It also requires listening to those with persuasive ideas. The act of cultivating relationships of connection, care, and community offers an urgent challenge to the extractivist status quo. This includes treating our lands and waters in a reciprocal way rather than as resources to extract. What would more caring relationships to each other and our environments look like?

These are huge questions. I do not have all the answers. What I offer is one counternarrative to the extractivist ideology that put us in this mess. Let's consider how to reframe the relationship to our environment from one of crisis to one of care. The lens I want to bring to this conversation and the necessary reframing comes from my current relationships and ways of thinking about feminist mothering and parenthood. My perspective about becoming a parent when faced with these ecological conditions stems from centring networks of community beyond the nuclear family, raising children with feminist ethics and values, and treating the environment — what some might refer to as Mother Earth — with reciprocity and respect.

From my vantage point, feminist parenting requires noticing our environments, inside and outside our lives, homes, and communities. While I was pregnant, two female colleagues and I formed a global network called the Feminist Environmental

Research Network — or FERN for short.[38] I see many parallels between this academic network and my personal approach to feminist mothering. Rather than supporting a competitive, ego-centric academic ethos, FERN centres community, relationships, collaboration, co-authorship, critical thinking, and creativity. These values are ones that I want to instill in my son as he grows into adulthood nurtured as he is by a critical ecofeminist home life. Research documents well how healthy environments produce healthy babies and healthy children. Parents know this better than anyone else. Many of the mothers I speak with in my various doula circles and mom groups share our fears about the future. Concretely, together in our local community, we share strategies, recipes, tips, clothing, baby gear, and book recommendations, and we go on walks, do yoga, offer childcare to help parents for date nights, and check in with each other periodically. Metaphorically, and physically, we cultivate networks of care and challenge the neoliberal myth that we can parent on our own. Isolation kills. I've seen far too many loved ones, friends, and family members struggle with mental health as they become new mothers, juggling the layered responsibilities and pressures of childcare, home organization, meal preparation, sexual satisfaction, creating a viable return-to-work timeline and managing shifting identities with a new sense of purpose. Many mothers feel they either fail at home or they fail at work. Through cultivating caring communities and supportive, community-focused feminist parenting, we can shift relationships so we don't always find ourselves alone, bearing the weight of the world on our own.[39]

My understanding of feminist mothering and parenthood is shaped by an urgent climate emergency and critical ecofeminist theory and practice. Each signals how we need to care about each other, the world, and future generations if we want to survive. Several years ago, I had the opportunity to contribute to an academic collection about motherhood, feminism, and politics. The excellent chapters set out an ethos of feminist motherhood that inspired further collaborative writing with mothers in the academy and informs my intellectual approach to scholarship as well as to mothering and to surviving the apocalypse.

Feminist parenting in my view means centring women and gender-diverse bodies, stories, experiences, voices, and perspectives in both health care policy conversations and climate emergency deliberations. It involves an ethos of care, love, and healthy relationships. This ethos, inspired by the late bell hooks, entails the challenging of patriarchal, male-dominant societies, promoting environmental awareness, centring partnerships, cultivating care, critiquing the individualistic ideology of neoliberalism, and addressing structural inequality.[40] Following bell hooks, feminism is for everybody, and it must be more than a lifestyle.[41] It requires intersectionality, a close investigation of the interlocking forces of race, gender, and class. Feminist parenting also draws attention to issues of reproductive justice. For future generations to flourish, to live with and envision life beyond the Code Red climate emergency, we need healthy environments inside our bodies, communities, and worlds.

───⋀─── CODE RED ───⋀───

Feminist Motherhood in a World on Fire

PREGNANT IN THE FIRST TRIMESTER, I struggled for breath. The unrelenting heat and wildfire smoke left me nauseous and trapped indoors. I couldn't see across the street. The intensity of burning wildfires is making it unsafe to be pregnant.[1] According to the BC Women's Hospital and Health Centre, "For pregnant people, exposure to short- and medium-term air pollution can result in decreased birth weight and possible risk of fetal malformations. Exposure to wildfires that require evacuation and relocation can also result in significant post-traumatic stress disorders."[2] With the layered crises of a global pandemic, wildfire smoke, the climate emergency, and the tender perinatal period, women's bodies experience unique vulnerabilities to climate change.

At the time, my partner and I were living in different cities, separated by the Rocky Mountains as he finished up his graduate degree. I was alone with our child budding inside of me, sick, and never more viscerally aware of how much the climate emergency felt gendered. Climate change requires thoughtful, multilayered intersectional analysis.[3] These "converging crises" require a multifaceted lens to investigate the gendered, racialized and class dimensions of extreme weather events.[4] It's not only women who are disproportionately affected; workers, especially many racialized and migrant workers, have few options but to breathe wildfire smoke and absorb extreme heat outdoors, with limited legal rights for recourse.[5] To be sure, it is an issue for humanity and more-than-human lives — humans, oceans, forests, animal migrations are all affected. I felt it in my guts, on my skin. In a poignant photo essay, sociologist Anelyse Weiler asks how scientists grapple with the emerging science about "how wildfires affect our lungs and heat waves affect our hearts."[6] There continue to be so many unknowns. Facing these uncertainties daily reflects the precarious hot mess atmosphere that adversely affects so many lives.

Pregnancy presents a unique sensory experience. My sense of smell heightened so much that if anyone dared to even turn on the toaster at home, I would yell out a stern warning about cooking anything except for bland veggie noodle soup. Even with the windows closed, in the fall of 2020, I could taste the wildfire smoke in the apartment.[7] I needed to get out. But where would I go? Gyms were closed. Outside was worse. The taste of carcinogens lingered in the air. I could not even *see* across the street. "Go to the mall," my partner suggested. Great, I thought. Go straight into the belly of the corporate capitalist beast. That's not going to save me, or humanity, from this all-consuming, burning mess. Grudgingly, I ended up walking around the mall in a haze, my mind as foggy as the external environment, lamenting the loss of *touch*. People came to the mall, trying to keep a six-foot distance, seeking sanctuary from the climate emergency with limited reprieve. Because I lived on Vancouver Island separated from family members by a lengthy ocean ferry commute and under public health stay-at-home orders, I hadn't seen my family in months. Alone and feeling all the more alienated, the absence of healthy familial touch felt palpable in the bustling mall, the atmosphere bland like the cold veggie sandwich I managed to choke down. Later that night, I listened to national and local news coverage, hearing that these environmental conditions would only get worse.

This kind of sensory awareness of one's intimate surroundings through smell, taste, sight, touch, sound, is one that I advocate for in my environmental teaching and scholarship, both inside and outside the classroom setting — encouraging citizens, decision-makers, and public officials to learn from lived experience and stories, to learn from affected parties, and move beyond linear forms of rationalist policymaking.[8] My lived experience of pregnancy during an intense wildfire season provoked further reflection on how the lived experiences of people with first-hand experiential knowledge of climate emergency events and their stories can inform policy and influence transformative change. On the brink of prenatal depression, with this all-consuming backdrop, I worried constantly about the future. And there were so many futures to worry about, most intimately, that of my unborn child, the future of my long-distance partnership, my health, the well-being of the students I was responsible for, and the future health of the planet.

From heat domes to atmospheric rivers and mass flooding — as I experienced in British Columbia during my son's first six months of life — a climate emergency forces us to reckon with the environment. These extreme weather events are matters of life and death, and they matter in nuanced, sensory ways, to all kinds of life. High heat can cause the body's organs to break down.[9] And the intensity of these environmental hazards will continue to increase with a fervent sense of urgency. Heatwaves. Droughts. Crop failures. River floods. Highway collapses. Mudslides. Erosion. Sea-level rise. Wildfires.[10] Forests more flammable than ever before. Not-for-profit groups like Sierra Club BC advocate for a shift away from heavy industrialized logging practices as a climate mitigation measure — given that old intact forests can moderate the landscape to support ecosystem function and lower the risk of the spread of wildfires, landslides, and flooding while protecting water sources due to their dense canopies, tough bark, extensive roots systems, and space among the trees and other growths.[11] This research reveals alternative strategies that can be employed in resource management to enhance ecosystems resilience. Such a paradigm shift is necessary given these exceptional circumstances, where unanticipated events have become the norm. The climate emergency is plunging us into survivor mode; these extreme events require new perspectives about living with this emergency climate, and preparing for what's to come.

Converging, multilayered crises reveal fundamental political tensions between capital and care. My experience of extreme heat was entwined with the global pandemic. When I was conducting research in 2023 with parents who were pregnant or nursing during the 2021 heat dome event, I heard several parents remark on the layered lived experiences of crises: of extreme heat, the pandemic, and post-partum depression, and limited social supports during these traumatic times.[12] Both the global pandemic and climate emergency force a collective confrontation with the politics of life and death. Such moments of crises draw attention to gaps in our social and environmental systems and incite new ways of thinking. As the pivotal thinkers behind the Hawai'i State Commission on the Staus of Women report *Building Bridges, Not Walking on Backs: A Feminist Economic Recovery Plan for COVID-19* discuss, along with political theorist Nancy Fraser on the contemporary crisis of care, women's voices and labours must be central to any viable plans for healthy and sustainable societal futures.[13] *Building Bridges* calls for paying urgent attention to the vital need to invest in social infrastructure: childcare, education and health care, which are lessons of relevance for Canada, too.[14] The voices of those directly affected by converging crises must be incorporated into recovery planning for healthier societies to flourish.

On one particularly sweltering day in the summer of 2021, I met up with a friend I hadn't seen since before the pandemic lockdown. We met during the early days of motherhood, when leaving the house felt like a rare occasion. We arranged to meet at a colleague's house with plans to go for a walk and eat lunch outdoors. Simple, I thought. Yet, with a newborn, even basic plans can turn like the wind.

Still figuring out how to dress as a nursing mom, I threw on some stretchy clothing and headed out the door. It turned out to be the wrong day to wear black. With sweat pouring and a fussy infant on my lap, I sat down in the garden to catch up with my friend who was accompanied by her new puppy and a colleague we both looked up to as a mentor. I fed my son two bottles of breastmilk, but his hunger cries son drowned out conversation. I panicked, sweat spilling across my brow as the sun bore down on us. I fidgeted in my lawn chair, nearly toppling over. I had gained 80 pounds in pregnancy. I had limited experience being physically vulnerable with colleagues — whom I usually saw in my comfortable campus office or the classroom — in such a condition of sweaty, sticky disarray. But on this day, I had no choice in the matter. My son was hungry, and I needed to feed him. Out came my breasts, one at a time, right then left. While my friend and colleague considerately ignored us, discussing politics and the state of the environment, I nursed my son, feeling completely unprepared and out of place.

Minutes that felt like eons passed. "Would you like a glass of water?" My colleague gently offered. By now we had surely overstayed our welcome. Sweat beads adorned my forehead before they streamed down toward my ravenous and presumably very dehydrated newborn. I accepted the glass of water, gulping it down. Thinking I'd embarrassed myself enough for one day, having had to bare it all in front of a senior male colleague, my friend and I excused ourselves and headed out on our walk. Only when I got home later that day did I realize that my shirt had been inside out the entire time. I sighed. Finally, when I got my son down for a nap later that afternoon, I curled up on the couch with a glass of ice water, exhausted, and turned on the news. Information about the heat dome — one of the deadliest extreme weather events we'd ever seen in British Columbia — dominated the screens. It seemed I would have to learn to live uncomfortably as a hot mess for the foreseeable future without much choice in the matter.

The year my son was born, global scientists and United Nations leaders called attention to a collective existential threat confronting all of humanity: the Code Red climate emergency[15] — an emergency unlike any other, moving both fast and slow. Fast in its manifestations of intensifying wildfires causing some communities like Lytton, BC, to burn to the ground on June 30, 2021, when winds of up to 70 kilometres an hour pushed a wildfire into the town. The mayor reported that a mere 15 minutes elapsed between people smelling smoke and the city going up in flames. Six months later, in November 2021, the province of BC declared its third State of Emergency of the year as extreme rain in the form of an atmospheric river triggered mudslides and blocked every major highway that connected the Lower Mainland of Vancouver to the rest of the country. Approximately 20,000 people were displaced, and thousands of animals died.[16] Other expressions of the climate emergency move slowly, like the rising seas creeping over coastal communities, engulfing and drowning out livelihoods around the globe. The Intergovernmental Panel on Climate Change (IPCC), a global body responsible for evaluating the science related to climate change, has warned that human influence is warming the planet at unprecedented rates.[17]

Facing the heat of one of the warmest summers on record around the globe in 2021, United Nations Secretary-General António Guterres called the extreme weather a "code red for humanity."[18] Just a few months later, during the 26th United Nations Climate Change Conference of the Parties (COP26) World Leaders Summit on climate change, Guterres continued his plea for increasing awareness of the intensity of the situation.[19] "Enough of killing ourselves with carbon," he cautioned. He then continued: "Enough of treating nature like a toilet." According to Guterres, when we burn, drill, and mine the earth, "we are digging our own graves." His stark warning calls attention to how the current climate emergency is a matter of life and death. By 2023, numerous communities around the world reported the hottest days recorded on Earth. Weather forecasters warn of exceptional warmth over a multi-year period due to emissions resulting from heat-trapping gases from burning oil, gas, and coal, as well as the return of El Niño, a cyclical weather pattern.[20]

We feel this matter in our flesh. Simultaneously, this global UN governing body becomes a harbinger of how all our bodies are intimately affected by these global events. Research shows how elevated temperatures coincide with increased emergency room visits for children.[21] Less is known about the effects of extreme heat on infants. We do know that extreme heat has long-lasting mental health effects, driving people to emergency rooms for care.[22] The climate emergency is a public health emergency.

Climate change affects all stages of gestation for mothers around the world. Some research now documents a correlation between premature births and our intensely changing climate, with particularly dire statistics for communities of colour.[23] The risk of a preterm birth is significantly greater for Black women, who are 50 percent more likely to give birth early than white women.[24] Air pollution is also a factor that has an impact on babies in utero. Studies show that if babies in utero are exposed to air pollution, then they may be born prematurely, stillborn, or underweight.[25] These studies reveal how climate change affects underprivileged communities unevenly compared to the general population when it comes to birth experiences and patterns. Higher temperatures and heat waves are also associated with premature births.

Three cascading State of Emergency events in British Columbia — the June 2021 heat dome, the burning of the town of Lytton, BC, that month, and the November floods — alerted people locally and globally about the immanence of the climate emergency. As new mothers, parents, and family members, we are living through this, charged with the responsibility of cultivating conditions for sustainable futures from the core of our being immediately and for generations to come. This provokes a radical reorientation of how humans relate to more-than-human environments and requires new ways of being together, not just thinking about but *caring* about one another and our environments. The only way out of this hot mess is to centre care rather than rely on extractive relations to each other and our environments.[26]

There is a dire need for a diverse array of voices in climate conversations. Women's voices and the voices of queer, gender-diverse communities cannot be excluded. Instead, they must be central. Those most directly affected, including women, women of colour, queer families, poor working-class families, migrant workers — all those invested in cultivating new families and relations — have direct lived experiences of the links between healthy external and internal environments for reproduction. An array of voices must be heard loud and clear in contemporary climate conversations. Unfortunately, those making important high-level decisions about global climate futures are often executive elites and leaders of national governments around the world, and thus, often exclude marginalized voices.[27] Yet, it is now widely documented that women often vocalize their lived experiences while living at the front lines of climate change–related movements, leading non-profits, and calling for governments to be held accountable.

Vital to shifting critical policy debates about how to cultivate more caring human/more-than-human relations for future generations is having the right people at the centres of decision-making authority. Shifting perspectives, discourses, and narratives is pivotal to this. Young people, women, parents, elders, and Indigenous leaders represent some of the voices all too often sidelined by heads of state. Yet their lived experiences and knowledges of human/environment relationships are essential for alternative, healthier futures.

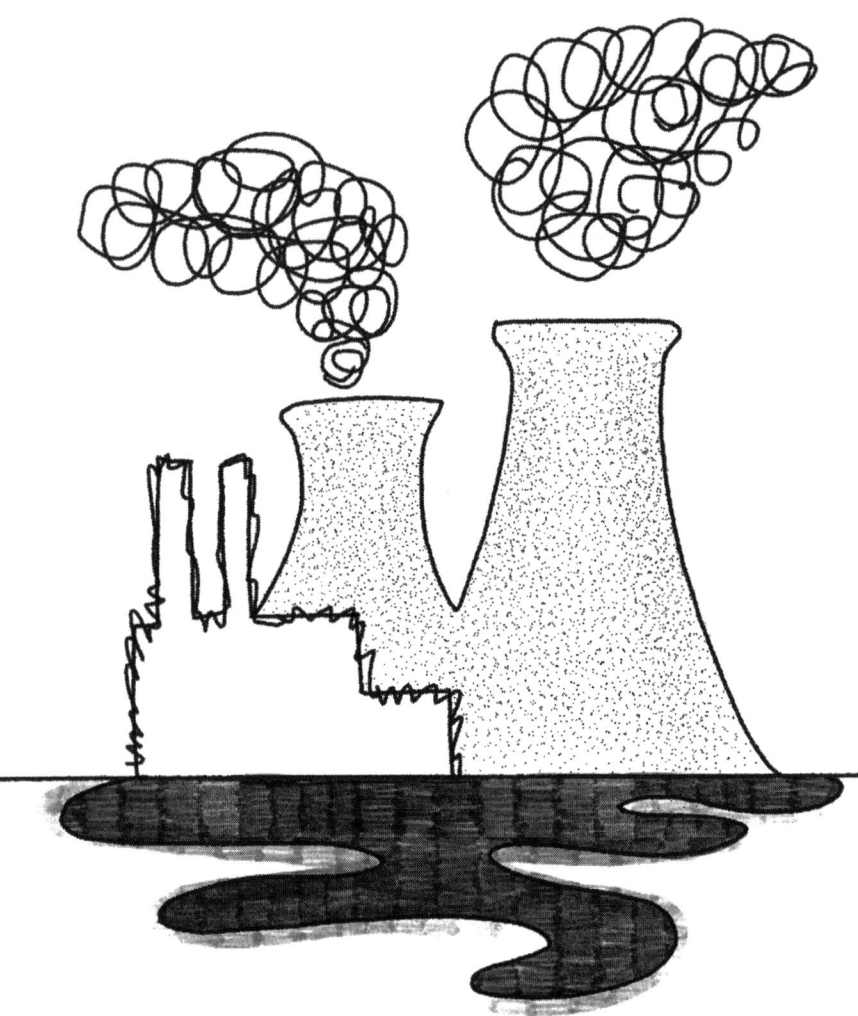

───⋀⋁⋀─── **CODE ORANGE** ───⋀⋁⋀───

Cultivating Community through Disasters

IF YOU TIME IT JUST RIGHT, as you approach Canada's Chemical Valley at sunset, a warm amber glow welcomes you to the city outskirts. This orange hue animates the landscape along Highway 402, a thoroughfare at the edge of Sarnia, Ontario. This artery frequently routes truck drivers straight across the northern edge of Chemical Valley loaded with materials, headed across the Bluewater Bridge, over to the Canada-US border, and then on to Detroit, Michigan. Those passing by may not even notice the dense concentration of dozens of smokestacks south of the highway. It takes effort to look and to notice. It takes even more effort to learn about what life is like living amid the living legacy of Canada's industrial, colonial empire, a region that has profoundly impacted the reproductive futures, physical, ecological, and cultural, for the Aamjiwnaang Nation.[1]

With street names like Imperial Avenue and Confederation Street, the city of Sarnia has a lengthy and layered geopolitical history that connects its foundation to the discovery of oil in the region. Sarnia's polymer refinery once produced rubber to bolster the global Second World War effort.[2] To commemorate these contributions, this image graced Canada's $10 bill in the 1970s and 1980s. Over the years, an assemblage of petrochemical and polymer refineries came to dominate the landscape and occupy the territory, due to controversial land sale agreements and the legacy of colonial laws that infringed on Indigenous self-determination rights.[3] Multinational corporations from Shell to Suncor have deep ties there, with subterranean roots that connect these plants to each other for the processing of chemicals and materials ranging from batteries to chewing gum. There is more than meets the eye in Chemical Valley, as these subterranean systems lie deep underground. Locals tell stories of industrial developers encountering human remains. Then life carries on, for both the local residents and the corporations.

If you pause to fuel up, as I often did during my many research site visits to Aamjiwnaang — an Indigenous reserve sandwiched in the middle of these chemical refineries — you might end up on the south side of the highway at an unassuming diner, the Leaky Tank. Local Aamjiwnaang residents have a sense of humour that is at once dry, dark, and colourful, holding together many complex emotions. A deep love for this polluted place they call home pairs with fear about the future. Every day, with their bodies exposed, the Indigenous residents of Aamjiwnaang breathe in an unknown mixture of carcinogenic chemicals, ranging from mercury to volatile organic compounds. Day and night, the smokestacks shoot effluent up into the air. Orange flares burn bright. Concerns about the health of their culture, community, and environment weigh heavily on local residents' hearts and minds.

On my very first visit to Aamjiwnaang as a young doctoral student at the University of Ottawa, I met with Ada Lockridge, a band member and mother, over a meal at the Leaky Tank. I wasn't the first outsider she had met there. Ada was, and continues to be, a tireless advocate for her community, meeting with journalists, scientists, lawyers, and policymakers to show what life is like for those whose lives are intimately entangled with Chemical Valley. The band named the local air monitoring station next to the band office in her honour to commemorate and acknowledge her sustained, heroic efforts to hold industry and government officials to account, noting her advocacy as the "community face of environmental protection."[4] At the time of our meeting, Ada held multiple official and unofficial leadership roles. In addition to her status as a band member living on the reserve with a view of Shell from her porch, and being a busy mother, she served as the Vice Chair of a nonprofit group called Victims of Chemical Valley — a group mainly comprised of widows whose partners passed away after working for years in the refineries due to health and safety concerns, and illnesses such as mesothelioma. [5] Ada was a founding member of the Aamjiwnaang Environment Committee. She also participated in a slow-moving, underfunded, and controversial health study, the Lambton Community Health Study. Over coffee, eggs, and toast, we discussed her decades-long activism, what kept her going, the various health and environmental studies she had participated in, and why she held out hope for change.

For the rest of the day, and the days following, with the scent of sulphur in the surrounding atmosphere, I couldn't get the taste — or rather stench — of rotten eggs out of my mouth. After breakfast, she took me on what she called a "toxic tour." On the tour around the Aamjiwnaang reserve, she showed me sites of cultural, environmental, and political significance, all within the backdrop of Chemical Valley. She joked that she should give visitors scratch-and-sniff cards to take home with them as a reminder of the various chemical smells residents live with every day. I saw polluted ponds, rivers, and forests. The baseball diamond next to the band administration office is separated from the plants across the street only by a highway and a chain-link fence, a stone's throw, or rather, a baseball pitch away. I saw people ignoring TOXIC warnings and KEEP OUT signs while fishing. The last stop on the tour was the cemetery.

Ada explained the many multijurisdictional layers of policy relating to her community's well-being, including band council, municipal, provincial, national, and international decisions that contributed to the devastation of this community.[6] I could feel, and almost taste, these policy decisions in my body, my gut.

Suddenly, the wail of a siren pierced our conversation, making me jump. "What the hell is that?" I asked Ada. I couldn't hear her response over the noise. Noting her calm composure, I looked around, scanning the perimeter of the cemetery, watching the orange windsocks blowing in the wind just over another chain-link fence. Having waited out the seemingly interminable siren for 15 seconds, she explained that the chemical alert sirens are tested every Monday at 12:15 p.m. sharp. If they sound any time other than on a Monday at this time, residents must shelter in place, or go inside, close the windows, and turn on the radio. After a few minutes, my heart resumed its normal beat. I couldn't help but think about how this was not a place one easily rests in peace. The rotten egg smell made my stomach turn and would stay with me for days after. Tears pricked my eyes and then flowed from the sting in the air and the realization that the development of Chemical Valley in and around Aamjiwnaang did not happen slowly or by accident, but by design.

After leaving the cemetery, we sat quietly. Ada registered my silence and offered me a coffee. We drove north to the Tim Hortons on Vidal Street, which would become a familiar destination over the duration of my long-term research relationship with Chemical Valley and residents of Aamjiwnaang, which began partway through my doctoral studies, now well over a decade ago. I wondered if I'd have the stomach for this research.

I felt disgust for the policy assemblage for environmental permitting and toxic exposure enabling the chemical plants to consume the community's lifeblood. Slowly sipping mint tea, I reflected inward, wondering whether I could even contribute in any meaningful way to amplify the voices of community members like Ada, whose stories needed to reach much higher levels of authority, and sustained intrigue as I wanted to know what made this place so special, why residents continued to reside here, defiant and strong against the encroaching industry. We sat in the lot for a few minutes, watching the local workers — many clad in bright blue jumpsuits — duck in and out on lunch break.

"This concludes the tour," Ada said to me. She drove me back to the Leaky Tank, where my rental car was parked. There was a lot to process on my long drive back to Ottawa. I promised her that I would be in touch as I wrote and prepared to defend my research proposal. Ada had agreed to discuss the ideas and be present during the defence, something unusual within my university setting, but which felt important given the community-driven context for the project. "I'll keep you in line," she promised with a smile as we parted ways, each of us appreciating the offer to continue our conversation. Over the coming years, Ada would be a mentor to me, someone I confided in, the person whose advice I took to heart even when it hurt.

This kind of project required time to build relationships and to treat communities as more than objects of study or investigation. I realized that I would not be able to stick to the library or archives to do this work, but needed to speak with people, conduct interviews, and hear life stories from those willing to share. When I returned to Ottawa, one of my professors said to me: "Sarah, your work is so emotional." I was annoyed and outraged by this remark at first, worried that he didn't consider my research to be serious scholarship. "Of course it is," I responded. "If you are going to spend several years investigating a topic of interest, you have to care about it. Otherwise, what's the point?" He agreed, while acknowledging that this kind of intimate, close, and personal research deviates from the norm in political science, a discipline dominated by cold, removed calculations and predictions of institutional norms and patterns.

One night, I caught a documentary produced by the CBC (Canadian Broadcasting Corporation, Canada's public broadcaster) called *The Disappearing Male*, about the effects of toxic environments on reproductive health. Ada was one of the mothers featured in the documentary voicing concern about reproductive health in Aamjiwnaang and the surrounding Lambton County. In 2005, Ada teamed up with scientists and documented a skewed birth ratio between 1984 and 2003 that saw more girls born than boys.[7] Their community-based participatory study showed a sharp decline in live male births starting in the 1990s. Ada and Aamjiwnaang community members wanted to know why, fearing this skewed ratio might have something to do with the unknown mixture of chemicals in their local environment. The study was inconclusive, calling for further research to link reproductive health patterns to toxins in their atmosphere, waterways, homes, and bodies. Though many studies have been conducted, Aamjiwnaang has seen limited concrete, transformative change to the environmental permitting process in Chemical Valley that also takes into consideration planetary health and cumulative impacts of toxins in the surrounding atmosphere. Yet, the community remains steadfast, holding their ground in this place they love and that they have called home since time immemorial, long before the founding of Canada.

As a political scientist, I was interested in how the community organized to document and articulate their concerns, how the various levels of government responded, and how this issue was framed by a range of actors. In my book, *Everyday Exposure: Indigenous Mobilization and Environmental Justice in Canada's Chemical Valley,* I discuss these layered policy configurations to examine the power relations embedded within institutions, discourses, and practices.[8] My book was largely inspired by the perspectives and worldviews of Aamjiwnaang community members as well as the scholarship of Rob Nixon, who discusses a kind of *slow violence*, which occurs so slowly that it is often not considered to be violence at all.[9] Similarly, in Canada's Chemical Valley, Aamjiwnaang residents are affected day and night, in their environments and in their bodies by their location that is besieged by heavy industry. More recently, global studies report that the noxious odours can cause mental health ailments.[10] They can ruin lives and livelihoods. Smells are often regulated locally, at the city level, rather than the higher levels of provincial or federal jurisdiction. The governance of toxic chemicals in atmospheres, bodies, homes, lands, and waters operates within a murky, patchy, and asymmetrical policy assemblage, with residents frequently bearing the burden of proof of exposure yet carrying limited decision-making authority, Indigenous laws and treaties frequently overlooked. Too often, smelly industries concentrate themselves in low-income communities. An environmental justice lens unearths these uneven, unjust relations among public officials, corporate authorities, and affected communities, and aims to amplify the voices and stories of those living with and resisting these realities first-hand.

Code Orange is a hospital code used to refer to mass casualties, or natural disasters. In this vignette, it also applies to the context of slow-moving violence or disasters. The slow-moving, invisible exposure to chemicals in the atmosphere can be understood as a kind of slow disaster, which is difficult to document, articulate, challenge, and change. The slow disaster of Chemical Valley's air pollution affects the reproductive futures of residents on a daily basis and is a direct result of extractive capitalism and colonialism. A slow disaster can be contrasted with a fast disaster, which is often highly visible and spectacular, like a wildfire or flood. These fast disasters are often understood as "natural disasters"; yet, as many researchers argue, they would not have been so disastrous if governments had been better prepared. In fact, we might even understand these as manufactured disasters, or biopolitical disasters, affecting human and more-than-human life, both quickly and over time.[11] Because of the recurring patterns and persistence of chemical accidents, leaks, and spills, we can understand many of the lived experiences in Aamjiwnaang emerging in response to this kind of slow disaster.

The act of addressing the latent crisis of slow disasters requires an ethic of care. This relates to transforming economic relations, research relations, and relations with broader more-than-human environments. As I elaborate in *Everyday Exposure*, doing intimate research requires a certain open-hearted rapport, a commitment of one's entire body, mind, and spirit. Shortly after I met Ada, I was invited to a community sweat ceremony, initiated by the youth council and led by an Elder. My brother was also with me that night. After we emerged, he said he felt reborn. It was then that I felt a stirring in my gut, a strong feeling — what I would now call intuition — that in order to do this research in a caring, thoughtful, sensitive, relational way, I would have to spend more time there.

A few months later, I packed up my apartment in Ottawa and relocated to Sarnia. At the time, I told myself it would just be for a year, during which time I could better understand the patterns of chemical spills, leaks, and releases, interview community members and policymakers about their lived experiences, and build bridges between the community, academic circles and the confusing policy arena for Indigenous environmental justice. While living in Sarnia, I attended public town hall meetings for the Lambton Community Health Study, as well as board meetings; interviewed public representatives on the board as well as community members; and reviewed studies and literature related to environmental and reproductive justice with a focus on Indigenous community well-being. I was not just a participant observer, but an involved, observing, and engaged participant who came to care deeply about those who welcomed me so closely into the community.

During my time in Sarnia, I linked my cell phone and email to the local alert system so that I would be informed of any spills or releases. Many times, I heard the sirens wail, not just on Mondays. Sometimes they went off by accident. This didn't inspire confidence. With community leaders, I wrote op-eds to the local paper expressing shared concern about this confusing emergency alert system asking: *What does the emergency system mean?* The question seemed at once immediate and metaphysical.

I ended up staying in Sarnia closer to two years after being asked to work with youth leaders on a documentary film to show their peers their environmental concerns and to share knowledge about Indigenous culture. We coproduced a film called *Indian Givers* with the support of the band's education department, youth council, high school, and school board. More than a decade later, the lead narrator of the film went to film school to continue to share his truths with audiences beyond Chemical Valley, highlighting both the resurgence of Anishinabek governance and of culture along with the harmful impacts of life in a pollution hotspot.

The decision to move to Sarnia and leave the comfort of Ottawa sparked some controversy. Academic peers thought I'd lost my mind. I was asked: Why would you move to a pollution hotspot? Friends sent care packages. Rumours circulated: She's fallen in love. She's queer. She's never going to finish her PhD. Trying to date in Chemical Valley could be the subject of an entire other book, or perhaps a TV special.

Years later, at a book launch event in Victoria after I'd moved home to the West Coast, a dear friend and Indigenous scholar said to the crowd: "What I love about Sarah is that whatever she does, whether she knows where she is headed or not, she gives it her whole heart." My cheeks flushed at the comment, embarrassed, but she was right. It can be uncomfortable to talk about care, intimacy, and love in academia, a field where we overemphasize reason, rationality, and logical modes of analysis and inquiry. This approach made my academic committee and peers uncomfortable.

Yet, much Indigenous scholarship on research ethics and methods emphasizes building relationships. All too often, academics approach communities looking for information. Frequently, they enter communities, collect data, leave, write up reports, and publish their findings. As elaborated in academic-activist scholarship that emphasizes "research as resistance" and solidarity-building with Indigenous struggles for justice, many Indigenous and allied scholars challenge this extractivist approach, founded on taking from communities.[12] Instead, they call for a relational, relationship-focused approach, centred on care, community, and reciprocity. Instead of taking from a community, an ethical researcher will think about what they can reciprocally give to the community they are working with. Co-creation is central to this more collaborative approach. A relational researcher will not just think about how they make their mark on the academic landscape with a single-authored article or book published in their name, but consider outputs of value to communities. For example, rather than solely publishing in expensive textbooks or journals hidden behind paywalls, open access resources, zines, photo essays, op-eds, reports, interactive websites, and music are all creative avenues for the meaningful co-production of vital community stories. Decolonial research requires a caring ethic to approach communities as sites of learning rather than sites of knowledge extraction.

Now as a new mother, I read everything I can get my hands on about balancing care and work. Many essays I've read about being a mother in the academy draw attention to the issue of a rigid binary that continues to exist, a false separation between mind and body. Academia, surely, is considered to be the domain of the mind. Becoming pregnant, giving birth, and being a new mom are some of the most embodied experiences I can recall having. While going through each of these, I've continued to be an academic, teaching while pregnant, then giving birth and becoming a new mother while continuing to follow-up on projects, attend meetings, and write. There is no mind/body separation. The French philosopher René Descartes got it wrong when he said "I think therefore I am," though his message resonates across most of Western society. Instead, in alignment with recent tenets of feminist and critical ecofeminist thought, I find the words of Dutch philosopher Baruch Spinoza much more compelling: "I feel therefore I am." Affect, emotions, and feelings are vital to human existence.[13] Academic scholarship beyond critical, environmental, and feminist circles is only just starting to acknowledge the problems with dualistic thought. While affect scholarship continues to take flight, the practice of blending intellectual capacity with bodily lived experience lags in the academy.[14] Becoming a community-engaged researcher prepared me for becoming a mother, because it blurs boundaries and challenges norms. This multifaceted existence is messy and does not always fit neatly into the academic parameters, walls, boardrooms, and classrooms set up with the privileged, white male body in mind.

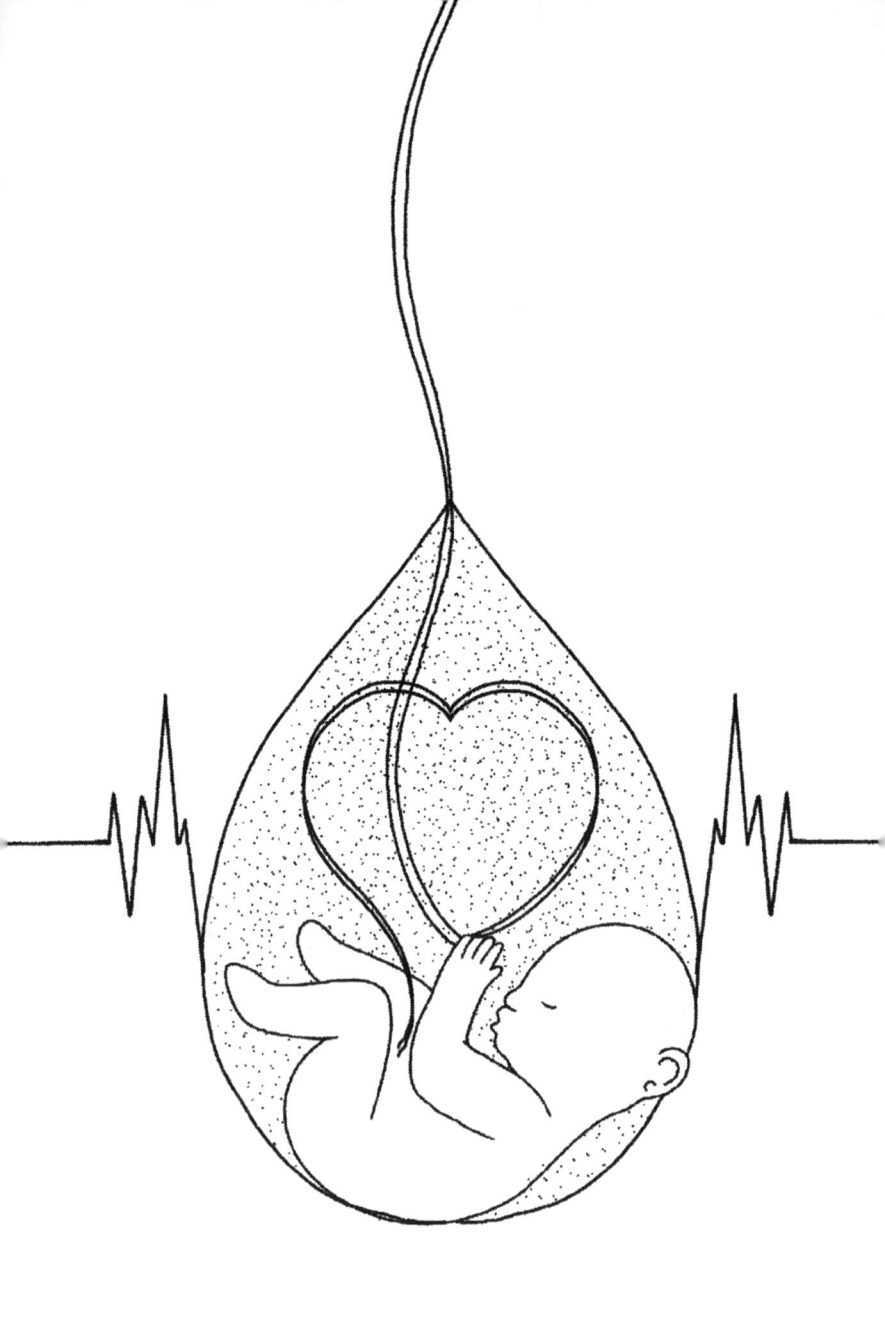

─── CODE PINK ───

A Splash and Dash Caesarean

I DIDN'T HAVE AN EASY PREGNANCY. I often felt like I was a nauseous, sweaty hot mess; at the same time, I felt an internal sense of calm and a duty to protect what my midwife often referred to as a "chill baby," due to their steady, even heartbeat. While I couldn't control the climate outside the womb, I felt a responsibility to raise a thoughtful, eco-conscious, feminist child while supporting their entry into an uncertain world. Pregnancy and the entire perinatal period of gestation brought out a genuine mixture of emotions: fear about bringing a new being into the world when faced with a climate emergency and excitement to welcome a new family member. Struggling with these tensions, with the question of how to raise a new generation when faced with a climate emergency and a world on fire, is a fraught experience addressed in the CBC documentary *The Climate Baby Dilemma*.

Giving birth was simultaneously the most traumatic and joyful event of my life. Indigenous governance professor

Dawn Smith discusses Nuu-chah-nulth teachings about yaʔakmis — the relational, entwined emotions of love and pain.[1] In her dissertation, Smith presents love and pain as a relationship between possibilities and limitations. While she discusses this journey in the context of postsecondary education and challenging colonial ways of knowing, I found solace in this concept while processing the grief of a challenging birth and bringing an infant into a burning world.[2] Following from Nuu-chah-nulth scholar Richard Umeek Atleo, Smith elaborates how yaʔakmis is a concept about holding multiple emotions in tension.[3] Becoming a parent certainly means holding together many complex emotions that are often in tension with one another simultaneously.

It can be hard enough to care for oneself during a warming world, and now we find ourselves charged with caring for and nurturing the well-being of future generations. Giving birth brought this into sharp focus for me. While pursuing research on the topic of extreme heat exposure and the perinatal experience in British Columbia, during several sharing circles, mothers brought up repeatedly how the layered crises of the pandemic, climate emergency, and postpartum depression were taking a toll on maternal mental health and well-being. Celebrating the joy of birth while acknowledging the pain and trauma required to bring my son into the world feels parallel to the mess of a world humans have created and how we must stick together and enact more caring relations otherwise.

During the early stages of pregnancy, I was debilitatingly nauseous and sick to my stomach at all hours of the day and night. I had to reschedule classes, cancel meetings, turn off the video during numerous virtual calls, and take frequent naps. More than once, I vomitted on the street, in the car, and even in the background while my partner was simultaneously delivering a presentation in real time. The pregnancy "glow" evaded me; I was stuck with something more like a haze, taking a steady dose of Diclectin and ginger ale to combat the nausea while gestating another being. I ran hot and struggled to keep my cool. My body of expanding organs built a placenta as my uterus grew. The fetal-maternal exchange dance took over and I was no longer in control of how I felt, managed time, or nourished myself. Becoming a mother meant letting go of some elements of control and adapting to a totally new tempo.

While pregnant and living in Victoria, BC, Coast Salish territory, the wildfire haze from California blew north. The past six years have been the hottest on Earth. We are living in a world of warmer, more violent weather.[4] Confronting this global reality of a "world on fire" from the solace of my apartment, I was alone, often struggling for breath, separated from family and loved ones, trying to construct a new home and a new life emerging inside me.[5] My therapist reminded me that I was not alone, that I was cellularly connected to this budding life inside me. Akwesasne Elder Katsi Cook informs us how "Women are the First Environment. In pregnancy, our bodies sustain life. At the breast of women, the generations are nourished. In this way, we as women are Earth."[6] Her words now carried new meaning in my life.

In the fall, my partner and I received our first sonogram reading. We heard the baby's beating heart. For the remainder of the pregnancy, the baby's vital signs, we were told, were strong, steady, and healthy — nothing to be worried about. In the meantime, we embarked on a housing search to accommodate our growing family, moved, nested, received many hand-me-downs, gathered used baby items, including a stroller and bassinet, and organized a virtual baby shower. My partner and I joined a weekly online childbirth preparation class. I went to frequent appointments with our midwife, read countless books and downloaded several pregnancy apps, discussed birth plan preferences. We decided on a hospital birth in case any complications arose, forgoing our plans for an outdoor water birth to avoid an emergency transfer to the hospital. I wanted to arrive early at the hospital and establish the atmosphere that would set me up for an unmedicated birth. This was Plan A.

Once the third trimester came around, my mobility became much more limited. My midwife and doula offered tips on how to ripen my cervix and prepare my pelvic floor. My partner often joked that I was born for labour; nobody in my family, neither my mother nor my sisters delivered their babies with medications or surgery. I played rugby for years, am covered in tattoos, and generally have a high tolerance for pain. When my contractions start, I would be ready, I thought.

Canada's paid maternity and parental leave had compelled me to leave a tenure-track position at the University of Hawai'i, Mānoa, and return home. I arranged for maternity leave to begin a few days before my scheduled due date. The baby's due date coincided with my partner's final assignments for his graduate degree. We held our breath and hoped the little one would arrive late. Earth Day would be perfectly appropriate and fitting, we thought.

Our baby's due date came and went. The only sign of labour was the popping of my cervix cork minutes after my partner completed his final presentation. What a considerate baby to wait until that moment, we amusedly told ourselves. Then Earth Day passed. We caught up on sleep, resting and nesting, hoping to bank this downtime for the postnatal period of what was sure to be incessant sleepless nights. Victoria General Hospital scheduled a few "non-stress tests" — the name of which inevitably stressed me out — to check up on his fetal development. Everything looked good, we were told. A week after the due date, we did another ultrasound and learned that I had a uniquely shaped placenta: a bilobed placenta, formed in two parts, like Mickey Mouse ears; not only that, but it was also nearing its expiration date. A few days later, we returned to the hospital for an OB-GYN consult. "We do not advise leaving this for more than a few days," she told us. I was cutting it close to the 42-week mark, and the obstetrician scheduled me for an induction the following Monday, my birthday, nearly two weeks beyond the baby's initial due date.

Luckily, contractions started just after the OB-GYN consult, and active labour began that evening. We slept with our eyes open. Around 5:00 a.m., I insisted that we go to the hospital. We called the birth support team — a midwife, her student, and my doula — to let them know. It was the most painful drive of my life as every bump and turn enhanced the pain of each contraction. We arrived before the hospital doors were open and buzzed in for support. I sat down on a wheelchair that resembled a shopping cart, as my partner wheeled me to the third floor and the hospital staff carried our carefully prepared hospital bags. I was in too much pain to set up the atmosphere I wanted. I was six centimetres dilated and set up camp on the bed instead, from which I didn't leave for hours. My water broke. Time became fluid.

For eight hours, having tried a variety of positions and spending hours in the hospital shower, the baby's heart rate remained strong and my contractions consistent, but I wasn't dilating past 7 centimetres. I was in the zone, breathing and roaring with each contraction. In 24 hours of labour, thanks to the support of my birth team and their hands-on pain management techniques, I made it without any medical interventions. Then everything changed.

I returned from a lengthy shower session for another cervix check. No progress. Upon the request of my midwife, an OB-GYN came in to assess the situation. She presented me with a menu of options to progress the labour and requested to put an IPUC — intrauterine pressure catheter — into my birth canal to get a reading on the force of my contractions. It didn't work. She shot up another device into my birth canal. I withered in pain as she fumbled around in the most sensitive part of my body, between contractions. She read the findings and recommended oxytocin to boost the intensity of the contractions and keep me on a path to a vaginal birth. It also looked like the baby was stuck as my pelvis was not creating enough space for him to emerge on his own. I was terrified of enduring endless hours of intense labour, but agreed to the IV and oxytocin hit to move things along.

When the nurses hooked me up to an IV, suddenly the baby's heart rate dropped to 55, well below the normal range of 110–160 bpm, demonstrating significant fetal distress. My midwife pulled the emergency cord from the wall behind my bed. Soon health care professionals flooded the room. I was flipped side to side like a sausage in an attempt to reactivate the baby's heart rate, though I didn't know what they were doing or why at the time. The OB-GYN came back in and showed me the reading print out — the baby's heart rate was so low it wasn't even registering on the data chart. Code Pink: obstetric emergency. She said we had to move fast and perform an emergency C-section. I would be unconscious for the remainder of the delivery.

On our way out the door, shortly before one nurse banged my bed into the doorframe and after another inserted the IV crookedly into my wrist, the OB-GYN stood beside me and recited the potential risks: organ damage, internal bleeding, hemorrhaging, infection … as my mind raced to process the events swirling around me and over my body, I lost track, lost control, and felt like an immobile corpse being wheeled into the operating room just across the hall. The declaration seemed more like a statement of risk factors rather than a meaningful discussion about informed consent. My partner was asked to wait in another room.

There was no time for him to scrub in. No time for the staff to sanitize my body. No time to count the tools before commencing surgery.

I was put under anesthesia for 30 minutes but the surgery itself took approximately three minutes. While I was under, my blood was taken; the reading noted extremely elevated white blood cell levels, an indication of shock, stress, or fighting off infection. Before I came into consciousness, my body was X-rayed to see if any tools were left behind. I missed many moments, viewing the birth, hearing the baby's initial cry, holding him in my arms, the Apgar test results that show how well the baby tolerated the birthing process and adjustment to life outside the womb, determining what to do with the placenta. When I came around, I recall seeing my partner with his bare chest, holding our baby. I smiled, unable to speak. Grateful, confused, hurting, numb, and healing, all at once.

A Caesarean was not in my birth plan but it should have been; about one in three births end up in a C-section, and these numbers are on the rise.[7] Normally there is time to prepare the body, and the birthing parent remains conscious. Mine was a more exceptional undertaking, involving ventilation through a tube inserted in my mouth which caused pain for over a week. It hurt to swallow, sneeze, or laugh. The catheter inserted into my bladder meant I was bedridden for the first 24 hours. I swelled up to 30 pounds over my predelivery weight after giving birth. My incision was crooked, an indication of the speed of the surgical operation.

Emily Dickinson wrote, "Tell all the truth, but tell it slant," reminding me of how birth experiences may be tiring and joyful, and they are also traumatic and life-altering. My father, a doctor who used to deliver babies, called my delivery a matter of life and death. All birth experiences fall along this continuum to some degree. Mine felt uncomfortably close to the latter. He informed me that there are degrees of emergency C-sections performed; any unplanned C-section is deemed an emergency. Given the expediency, lack of agency, and loss of consciousness, mine was a "splash" or "crisis" emergency — sometimes called "splash and dash" Caesareans: the birthing body crashes, unconscious yet still contracting. Then, the medical team splashes some disinfectant on it and dashes into surgery.

Though rendered immobile, my body became a carrier of life. The scar tissue from the incision will forever remind me of how my body became a portal, symbolizing the entangled and related emotions of love and pain. Like the suture that heals an open wound, speaking with others — family, friends and therapists — and writing serve as cathartic forms of stitching together the pieces of this story. I retell this story to refocus the light and to honour the beautiful emergence of this new life while simultaneously contributing to a much-needed public conversation about the private affair of birth, including the layered complexities of informed consent, reproductive mental health, and the prevalence of emergency C-sections. Issues related to the birth experience seem too often lost in the shadows. By sharing this story, it is my aim that it will contribute to more robust conversations about these vital topics — matters of life and death.

There is not much time to process the trauma of birth while simultaneously being thrown into the role of a new parent. I was so fortunate that my partner was able to spend the next few nights by my side in a private hospital room, which we affectionately and deliriously called the hospital penthouse suite — where they serve prune cocktails on the menu, and a trip to the hospital shower was a luxurious indulgence compared to the pain of birth. As I was fed antibiotics and fluids through an IV on my very swollen left arm, and had my blood drawn periodically from my now incredibly bruised right arm, in addition to the recent surgery in my belly inhibiting free movement, I was bedridden at first. Once the catheter was removed, a nurse walked me to the bathroom as we wheeled the IV beside us.

As day bled into night, we found ourselves alone, trying to figure out how to care for our baby. At 4:00 a.m., we called in the nurse for help soothing our son to sleep: "Try patting his bum," she suggested from the doorway before quickly moving on to the next patient. Sleep escaped us that night.

One of the many nurses I met during this blurry period finally moved me from an IV to oral antibiotics so I could shower once again — much needed as the first since a very messy birth. My colostrum came in, so I could begin feeding our baby from my bed. The loss of sensation from the general anesthetic soon transformed into sensory overload as I began to feel and process the wounds from a complicated birth. Similar to what Tabitha Soren depicts in her photo essay for *The Atlantic,* "The Blurred Existence of Motherhood," our first day as parents was a blurry whirlwind of pain management, trying to squeeze in broken rest, learning how to keep our son alive, soothe his distress, and finally informing our families why we had been out of touch for more than 24 hours.[8]

To manage or muddle through, we learned various breastfeeding positions. Feeding led to a rush of oxytocin in my uterus, which created cramp-like pains as my uterus began its process of involution (where the uterus returns to its prepregnancy state) and the release of lochia (vaginal discharge). The expulsion of fluids was another focusing event in the room. From changing the baby's meconium-filled diapers to learning his rooting signs, the engorgement of my breasts and breathing through the sensations of diastasis recti (a postpartum condition that involves abdominal separation) as my body slowly began to return the organs and muscles into their prepregnancy place, I barely had a moment to process the twin emotions of love and pain enframing this entire experience. Thankfully, my body responded well to the pain medication. My white blood cell count dropped significantly enough that we could be discharged 48 hours after delivery, on my birthday. I insisted on indulging in the bittersweet taste of my first glass of champagne after nine months of abstinence.

The perinatal experience reveals the ways in which the pregnant body carries the weight of the world. This felt particularly acute in the context of giving birth and nursing during a Code Red climate emergency. In *The Cultural*

Politics of Emotion, feminist scholar Sara Ahmed discusses scar tissue and the politics of body sensations. The surface of skin is a place where emotions collide. Accordingly, "the scars on your skin both attach you to a past of loss and a future of survival."[9] Each time I look at my slanted scar, I think about the story of this birth and reflect on the beautiful emergence of my son. My body became a portal, an opening to give him life. An embodied symbol holding together the emotions of pain and joy, fear about raising a son amid a precarious ecological future, and hope for alternative possibilities beyond the status quo.

Many questions linger. We don't have all the answers, and most likely never will. What caused my son's rapid heart rate deceleration? Was my uniquely shaped two-pronged placenta to blame? Where did my placenta go? So much for my desire to ceremoniously send it off to sea, to carry on in the watery, oceanic environment I had envisioned and written into my birth plan. What matters more than answering these questions is creating space for conversations about complicated births, so that splash and dash emergencies are neither normalized nor silenced, and so that women do not feel afraid to discuss their birth experiences and entangled emotions of love and pain that transpire through all stages of birth.[10] Following Ahmed, I am optimistic that the shared language of pain can connect one body with other bodies to cultivate conversations and senses of community around these sensitive situations. By giving flesh to feelings, sharing and caring together through such intimate, visceral, and intense experiences, there exists space for healing. There is much we can learn from these scars. Every scar tells a story. Scars simultaneously remind us of painful injuries alongside the imaginative possibilities for other worlds and forms of life to emerge.

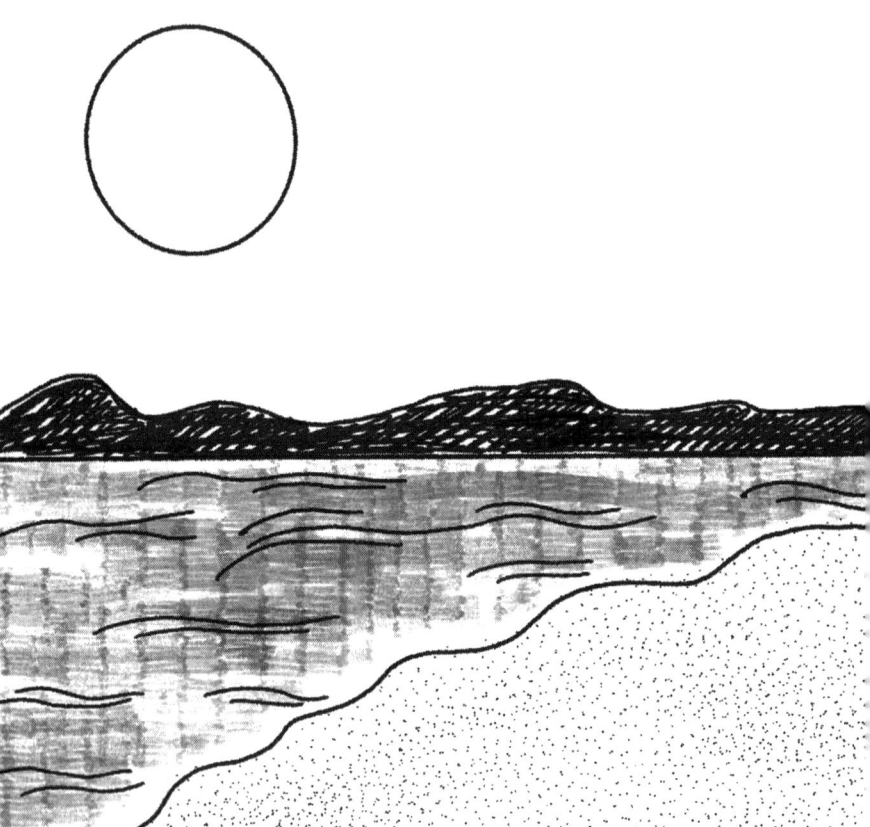

─────⋏─────── CODE BLUE ───────⋏─────

Living through Multiple Crises, Climate Anxiety, and Mental Health

FOR MANY NEW PARENTS, the postpartum experience is one of the most vulnerable times for mental health. This becomes an even more tender time during the current climate emergency. According to the US Environmental Protection Agency, pregnant bodies and their fetuses are more vulnerable to the health effects of climate change events ranging from extreme heat and flooding to wildfires, which, in turn, can cause health problems such as low birth weight and even miscarriage.[1] Moreover, pregnant and postpartum bodies can be at increased risk of post-traumatic stress disorder (PTSD) and depression after living through natural disasters and extreme weather events. Those who were pregnant, gave birth, or nursing in the lead up to and aftermath of the 2021 heat dome in British Columbia, for example, were exposed to a multifaceted triple crisis: adverse mental health, the climate emergency, and a global pandemic.

Extreme heat exposure is a matter of life and death. Mothers I spoke to during a 2023 research project on extreme heat raised the question of what it meant to be exposed to connected and converging multilayered crises, with the climate crisis alone unleashing fire, drought, extreme heat, and floods.[2] I started this research out of an interest in my own familial experience and a desire to co-create alternative policy configurations that would be more supportive during these tender and tenuous times. One mother explained that "when you're pregnant, you have so much to worry about already. And then with the heat dome and all these new weather events, it was a whole new layer of things to worry about." One mentioned "trying not to cry and overheat" during the heat dome. Most wanted more information about the impacts of extreme heat specifically and climate change generally on pregnancy and nursing bodies. Putting it into sharp perspective, one mother shared: "I feel like these are really, really huge, systemic problems that you know misting stations aren't going to fix." Some sought reprieve from the heat through the ocean, as I had done. For me, oceans have always been a source of strength and solace in times of personal strife. But this long-beloved source of comfort is now also another cause for concern.

Oceans are harbingers of planetary health. During the 2021 BC heat dome event, with shoreline temperatures rising well above 50°C, clams, mussels and sea stars died en masse, cooked by extreme heat exposure.[3] One billion marine animals died.[4] This harrowing experience signals the interconnections between bodies of water, governance of planetary health, and bodies of knowledge — the knowledge generated from both human and more-than-human life experience.

How individuals, communities, and marine ecosystems encounter the administration of policies and practices pertaining to managing the health of coastlines can be understood as, what I call, blue bureaucracy — the troubling governance regimes that assume human dominance and extraction over nature in coastal communities. The administration of blue bureaucracy takes shape through *institutional configurations* — policy decisions or indecisions; *discursive practices* — how we come to speak about and discuss our relations to waterways, oceans, or seascapes; and *embodied knowledges* — the multilayered ways in which we feel, sense, and live with rapidly changing coastlines and how they affect overall planetary health and well-being. Blue bureaucracies come into contestation with sadness, with how we come to experience, notice, and live with warming weather and shifting coastlines. Blue feelings — like heartache — reveal a tenuous connection between light and shadows, love and pain. Learning from the embodied knowledge of those directly affected by and resisting blue bureaucracies is one pathway toward transformative change. Through sharing stories about our experiences of crises generally and climate emergency events specifically, we can raise awareness, shift perspectives, intervene on problematic climate discourses, and influence climate-affirming policy outcomes.

Growing up close to the Pacific Ocean, at the edge of the Salish Sea on a body of water known as the "Indian Arm," I had an intimate relationship with seascapes. The ocean was a source of enjoyment, play, nourishment, and transportation. I was raised by the ebb and flow of the tides, volunteering many summers as a Beachkeeper in my neighbourhood. My childhood best friend and I would walk the shoreline and share information with park visitors about environmentally friendly beach behaviour. "Excuse me, are you aware of the protocols in place pertaining to crab sizes," we would often ask with an authoritative inflection that intimidated no one.

Coasts, always shifting, were constant places of learning and of intrigue. I learned to paddle a canoe and drive a boat before I learned to drive a car. I swam before I could walk. At the same time, it was hard not to feel blue — a sense of melancholy — about the erasure of the local Indigenous communities from this region. This is changing. The shoreline that grazes the edge of Belcarra Regional Park (across the bay from North Vancouver) was recently renamed təmtəmíxʷtən, to recognize the traditional territories — land and marine — of the Tsleil-Waututh peoples.[5] Yet, the name "Indian Arm" — the body of water along the shores of Belcarra and Tsleil-Waututh territory — sticks to this day, revealing how the ongoing legacy of colonial administrative authorities lacks local awareness about the diverse, vibrant, and rich cultural practices of Indigenous peoples, for whom these shorelines constituted their home for millenniums. Over the years, I spent many summers paddling up the edge of Burrard Inlet by canoe or kayak. To sail out to the Sunshine Coast, Gulf Islands or further into the Pacific Ocean, one must travel up the Burrard Inlet past Burnaby, under the Ironworkers Bridge, then on to the shores of Vancouver. Across from the beach is the terminus of Western Canada's main sources of infrastructure: the Trans Mountain Pipeline, also known as the Kinder Morgan pipeline expansion project.

For years, environmentalists and Indigenous Nations like the Tseil-Waututh have opposed the state's impetus to twin this pipeline that originates in Alberta's oil sands and crosses through British Columbia to export oil. A fierce commitment and dedication to the environment, a fervent love and passion, clashes with corporate interests and efforts to extract from vital oceanic life-giving forces. The ocean is a life force, not a resource, many Indigenous leaders and scholars have expressed over and over again.[6] Nations such as the Tsawout Nation, located at the edge of the Salish Sea adjacent to Sidney, a town 25 kilometres north of Victoria, presented oral and written testimony to articulate concerns about enhanced marine traffic and the impact it would have on their vital ecosystems, including threats to their fishing practices. During a formal public hearing in the fall of 2014, I witnessed testimony that an increase in tanker traffic would affect their self-determination, food systems, and coastal ways of life. This was a few months after I had paddled with members of the Tsawout Nation to document and co-produce a short documentary film, *To Fish as Formerly,* about how their relationships to this marine territory is a core part of their sovereign ways of life and their governance systems and vital to their being as coastal Indigenous peoples.

Oceans can also surprise us. In January 2022, submarine volcano Hunga Tonga-Hunga Ha'apai erupted, aftershocks lasting for more than 12 hours, sending sulphur dioxide into the air and engulfing the Pacific Tonga nation with ash; its momentum powered an ocean tsunami whose aftereffects were felt as far as on the West Coast of Canada, and whose reverberations made sound waves and atmospheric pressure "ring like a bell" across the Pacific Ocean.[7] A local ecological event echoed globally, transcending rigid international geopolitical boundaries. Tsunamis were recorded around the world including in the Atlantic, Caribbean, and Mediterranean regions.[8]

According to the National Collaborating Centre for Environmental Health, which tracks rising sea levels due to increasing glacial meltwaters and warming oceans in parts of Canada, sea levels may rise up to 175 centimetres by 2100, significantly poised to impact Atlantic Canada and British Columbia.[9] By 2050, US coastlines are set to rise by a foot, according to a National Oceanic and Atmospheric Administration study.[10] Such startling projections should be interpreted as an existential crisis, where climate change is a way of life, shaping seascapes, forcing erosion, flooding, storm surges, and relocation of populations.[11] Rising sea levels have major implications for beach erosion.[12] As physical geographer Jasper Knight explains, "Coastlines — the interface between land and sea — lie at the front line in the battle against climate change impacts" with an increase in climate hazards like floods, rising sea levels, hurricanes and tropical storms, and sands that erode easily due to waves.[13] This is especially troubling because sandy coastlines are significant for carbon storage and biodiversity. Features like estuaries, beaches, and sand dunes can buffer the adverse impacts of climate change.

Knight argues that it is essential to view coastlines as "green infrastructure" to manage harmful impacts. This paradigm shift requires thinking about coastlines not simply as a boundary between land and sea, but as being interconnected. What becomes of this vital public asset is a looming question mark. How can we envision the democratic life for human and more-than-human forms of life in these eroding conditions?

Coastlines are vital tidal places of life and death. For many communities around the world, the dramatic effects of climate change have led to displacement, including for over 2.5 million Americans.[14] The effects of climate displacement are felt globally but unevenly. For example, in Fiji, where coastal temperatures are rising, a cyclone is not just a theoretical concept but an experiential event — literally named "the wind that kills" or na cagilaba.[15] A cyclone shapes how beings live along and with their coastal environments, affecting everyday livelihoods and reducing time spent outdoors, away from activities of cultural and recreational significance, like playing rugby, swimming, and fishing. This is devastating for island communities like the archipelago of Fiji, one of the world's smallest contributors to global carbon emissions, yet facing such severe consequences and further illustrating the inequitable impacts of climate injustice.

In the fall of 2021, when my son was six months old, a large atmospheric river caused catastrophic flooding in British Columbia, prompting yet another State of Emergency declaration by officials across levels of government when a storm surge unleashed a month's worth of rain in two days on several communities across the province.[16] This third

State of Emergency in British Columbia was declared on November 17, 2021, with the COVID-19 global pandemic as a backdrop to these layered crises. An atmospheric river is a large stream of water vapour, generally with origins in tropical ocean regions near the equator.[17] Around Hawai'i, the Pineapple Express brings water vapour from the north across the Pacific Ocean. Atmospheric rivers can bring large masses of warm ocean water to the north, as they did when they pummelled the West Coast of Canada during a storm surge on Saturday, November 13, 2021, setting rainfall records across the province. Oceanographers around the globe agree that climate change is intensifying atmospheric rivers and making these exceptional extreme weather events the norm. Some scholars are calling for treating these events as systemic and allocating emergency preparation resources accordingly.[18]

An extreme weather event, such as an atmospheric river, is destabilizing to ecosystems, infrastructures, and community well-being. It drenches and unsettles mountainous terrain, causing landslides. Such fluid landslides can be provoked by wildfires, which incinerate the forest floor. An important barrier to erosion becomes lost. As night fell on Sunday, November 14, nearly every major highway across British Columbia closed. Eventually, Vancouver was cut off from the rest of the province. Thousands of travellers were trapped, stranded on highways under the deluge. Some travellers found themselves stuck in Hope, BC, with limited avenues out of the city. Farms and hundreds of thousands of livestock were submerged. Boats zipped across flooded farmlands to provide respite. Cattle were airlifted to safety. People flocked to grocery stores, emptying the shelves, worried about limited supply chains. Residents of the small towns of Princeton and Merritt took sanctuary on their roofs and waited for further assistance. Stormwater overwhelmed sewage systems as rain poured down and toxic water from the sewers submerged entire communities. Bridges and highways collapsed. The federal government called in the military. It is impossible not to recognize it as a climate emergency. Residents slowly returned after the waters receded, finding their homes caked in layers of mud and debris. Public officials called for gasoline rations as gas stations across the province ran dry and shuttered. Nonessential travel was banned.

The cost of repair to infrastructure alone will be more than one billion, we were told. Then consider the recovery costs borne by homeowners, businesses, and animals. But one cost remained obscured in public discussion. These extreme weather events function as the perfect storm for distress to mental health.[19] Researchers contributing to a 2023 IPCC report noted with high confidence how mental health challenges are associated with increasing temperatures, along with trauma from extreme weather events, and loss of livelihoods and culture.[20] One of the most obvious yet difficult issues to discuss during any disastrous scenario are the emotional ones. Blue feelings of love and loss, joy, beauty, and pain held together are emotions that linger, entwined.

Those living at the frontlines of compromised environments, or in toxic hot spots experience the mental health effects daily both consciously and unconsciously. The Waiʻanae coast of Oʻahu, Hawaiʻi has been overburdened with landfills, prompting the passage of Senate Bill 2386 to prevent landfills from further encroaching on the homes of Native Hawaiians, as the bill requires a half-mile "buffer zone" between the waste facilities and residential areas. Prior to the passing of the bill, local zoning ordinances allowed a disproportionate placement of hazardous waste facilities near communities.[21] Locals resist having their lands and waterways considered as dumping grounds, concerned that living next to toxic sites might cause asthma, cancer, or birth defects. The long-term effects of such proximity to hazardous waste are still poorly understood.[22] Similar concerns arise in Chemical Valley. The mental health implications of the layered concerns related to colonialism and environmental injustice are becoming more rigorously documented and researched. Leading critical geography scholar Farhana Sultana writes about "the unbearable heaviness of climate coloniality" and how the impacts of climate change are felt inequitably by communities who become vulnerable to the layered legacies of global racial capitalism, colonialism, and imperialism.[23] She elaborates how coloniality is experienced through both overt and covert ecological degradation in multiple forms — pollution, toxic waste, mining, disasters, deforestation, erosion, and so on. What we also know, as reported by the World Health Organization, is that healthy built environments are vital to ensure mental well-being.[24] Healthy homes, lands, and waters are crucial to healthy hearts and minds.

Living on the edge, at the coastline, provokes all kinds of feelings. Many crises are shaped by water.[25] It reflects. It becomes a portal. Because I previously lived in Hawai'i,

I received an alert message from Global Affairs Canada in 2021 with the following message:

> Storm in Hawaii: A storm has been affecting the Hawaiian archipelago since December 6, 2021. It brings heavy rains, wind gusts, electrical activity and snow in mountainous areas. The storm caused flash floods and power outages in some coastal regions, including Honolulu. It may also lead to landslides and damage to infrastructure.

I immediately checked the news, then checked in with people dear to me. These intense weather systems, just weeks apart, yet connected across the Pacific Ocean, weigh heavily.

For many in Hawai'i, life on the edge implies an everyday existence contending with their livelihoods under siege. Native Hawaiians live fighting for their sovereign rights daily in tandem with the gruelling realities of the intensifying effects of climate change. In 2021, the city of Honolulu issued a flash flood warning due to an intense seasonal cyclone — known as a kona low storm — bringing rainfall to areas that do not typically get much rain.[26] Given the vastness of the Pacific Ocean, it is rare for an extreme weather event to stall right over the Hawaiian archipelago.

Hawai'i Governor David Ige signed an emergency declaration on December 6, 2021, to allocate resources for the emergency response effort. The mayor of Hawai'i County declared a State of Emergency to leverage funding from state and federal authorities. The Hawai'i Emergency Management Agency urged residents to avoid nonessential travel. Businesses closed for the day due to power outages. People could not sleep to the sound of such intense rainfall. My friends and former colleagues from the University of Hawai'i, Mānoa, shared social media posts featuring heavy rainfall soundscapes accompanied by imagery of the deluge. One colleague circulated an email with an image of the view from her home on the "Windward" side of O'ahu. In the background, palm trees framed her view of the Pacific Ocean. In the foreground, puddles of water were filling on the streets and spilling over from the ridge of the shoreline. Ambulances were whizzing by. This is a fully immersive, sensory, embodied experience of climate change — of the current climate emergency. Then, Hawai'i again became vulnerable to climate change during the August 2023 wildfire that took out the town of Lahaina on Maui, with a death toll of 115 as of March 8, 2024.[27] The August 8 Lahaina wildfire is the deadliest to have occurred in the United States in over a century. This event illuminates how the climate emergency is an urgent matter of life and death. Those who survived continue to grapple with the loss of their homes, livelihoods, and loved ones.

Settings like these are increasingly the norm for frontline coastal communities, from my home in Coast Salish territory in BC to the shores of Hawai'i. Life during a climate emergency is inherently complicated, rife with blue feelings. I drove north along the windward side of O'ahu to the community of Ka'a'awa during the winter break, in the middle of the rainy season, then parked to watch the fishers cast their lines as the waves slammed their boats, setting them back, and poured over the highway, splashing my parked vehicle. To the left, I saw the stunning Ahupuaʻa ʻO Kahana State Park, which was the setting for many of the *Jurassic Park* films and *Jumanji*, and a popular horseback riding site for tourists. To my right, as I looked out to the horizon at the vastness of the sea, was an eroding shoreline. I found myself simultaneously grasping the beauty of the land and the sorrow of unrelenting climate change and its consequences.

But the communities living on these edges shouldn't be thought of as damaged. Even resilient isn't the right word for them. Viewing these communities as victims of an eroding coastline creates a superficial narrative that detracts from the causes of the climate emergency. Critical policy studies scholar Michael Orsini calls the use of the word "resilience" a "plastic term," the "positive residue left from trauma and tragedy" that permeates public discourse. It's a superficial narrative that masks systemic wounds and reinforces a kind of biopolitical, neoliberal emphasis on individual management of lifestyle choices, to borrow from the work of critical social theorists Nikolas Rose and Michel Foucault. Such plastic terms become void of meaning. And on the contrary, coastlines, and the people making life along them, are rich with abundant meaning.

We can take our cue from the stirring words of Pacific Islander poets to reframe such troubling narratives. Legal scholar and Guahan poet, Julian Aguon, argues that stories, not merely science, are necessary to win the fight against climate change.[28] In his moving essay, "To Hell with Drowning," he contends that the peoples of the Pacific Islands, those on the frontlines of Oceania, have powerful narratives to inform the contemporary climate justice movement. Their insights are born both of living closely with their local environments and of surviving the ravages of the extractive industry. Many communities living in the Pacific Islands, such as those in the Republic of the Marshall Islands, face the daunting reality of climate-induced relocation. For decades, the US military — the largest institutional producer of greenhouse gas emissions — has treated these islands as empty sites for testing ammunition, with little regard for the vitality of the peoples, the landscapes, the seascapes, and their connected ecosystems. This has resulted in loss of vegetation and a legacy of hazardous waste and toxic contamination felt acutely in places like the Pacific Ocean, including Kahoʻolawe off Maui.

Oceans and coastal communities are vibrant, dynamic, abundant, and full of human and more-than-human life. These places are anything but empty. Home to stories, seabirds, songs, traditions, and fishing practices, Oceania is not a vacant, empty space. As Tongan scholar Epeli Hauʻofa discusses in "Our Sea of Islands," islands in the region are not isolated atolls, but connected archipelagoes.[29] What does it take to reconceptualize how we look at our local environments, viewing them as sites of care, love, connection, and meaning rather than as sites of extraction?

In her visual poem, "Anointed," Marshallese climate activist, Kathy Jetñil-Kijiner, speaks about the deep love she has for wounded places in the Pacific, injured by nuclear testing.[30] She talks about the vibrant canoes, breadfruit, coconut trees, and beds of watermelon that animate what became a testing ground, a crater, an empty belly. Her words transform worlds through stories that go beyond ruinous narratives, and that are anointed with power. She talks about a people on fire, fighting back against the lies they have been fed about the health and safety of their communities. Her words demonstrate the generative power of the arts, grounded in community and storytelling, to speak back against extractive relations of empire. Many Indigenous leaders around the world centre the significace of water to the well-being of their communities. Tsleil-Waututh leader Reuben George's 2023 memoir *It Stops Here: Standing Up for Our Lands, Our Waters and Our People* is about the importance of ceremony to community wellness, culture, and healing.[31] His story informs readers how the sacred sweat lodge serves to replicate a womb environment, as a site of mutual care, connection, reciprocity, and well-being.

Turning to non-Western ways of being, knowing, and feeling in the world offers a source of inspiration and strength when living through these trying times. While much of classical Western thought has tended to isolate emotions from the body, female lived experience and emotions, Indigenous perspectives as elaborated here and in the words of leading activists like George instead centre the womb as a vibrant site of regenerative life. As such, grounding blue feelings in community and stories of strength offers an antidote to the climate emergency and to the corollary of eco-anxiety.

Motherhood itself can be isolating, more so during a global pandemic twinned with a climate emergency. The transition to parenting is not easy. Our social environments, built infrastructures, climate change, and affected ecosystems all play significant roles in our mental well-being — and require further investment. Mental health scholars and workers agree that this is an anxiety-inducing time.[32] We will be living with and analyzing the psychological fallout of the pandemic for years to come. When I fainted two years after spending yet another sleepless night, during an intensive teaching semester, after standing in the sun and not drinking enough water, resulting in a concussion and yet another trip to the emergency ward, I couldn't help but think this must be the result of these layered crises impacting my body from the inside out.

Our bodies are made of water, and babies come into being through this liquid environment. The health of these bodies of water are vital to the reproduction of vibrant and flourishing generations. Yet, women's bodies are understudied in all kinds of biomedical scenarios. The able-bodied, white male body remains the default data point. Since the inception of ancient and classical Western philosophy, the life-giving body has been sequestered to the "natural," unruly, messy, hysterical realm consciously separated from the rational, political arena of reason that is set aside for male intellect. There is no other event more embodied than pregnancy, and giving birth makes one realize how false this personal/political binary is and just how powerful and generative the body can be. Further, in the wake of repeated climate emergency events, rigid boundaries between the personal, professional, and planetary aspects of life are collapsing. Climate emergency events have visceral mental health effects, and call attention to the need to collectively grieve and come together during traumatic emergencies, both the fast and the slow moving.

As a new mother myself during these layered emergency events, I became intimately familiar with emergency ward settings. First, to give birth. Then a few months later, during the tail end of the heat dome and unrelenting heat wave that scorched Western Canada. And then again two years later during an intensive period of toddler and student care, when I fainted. As a newly breastfeeding mother, I could not keep up with the needs of my nursing son and fainted. We revisited the emergency ward again six months later, when we discovered my son's egg allergy and once again experienced triage, care, and an ambiguous treatment plan — the taste of eggs lingering and our own breakfast plates unfinished. Living within these layered states of emergency — pandemic, Caesarean, heat dome, storms, the climate emergency — are disasters not "out there" affecting others far away, but viscerally hitting home.

Helping my son develop a sense of agency about the urgency of the climate emergency without leaving him feeling disengaged or disempowered is going to be a significant area of communication in our family, his circles, and for future generations. As he grows into his childhood and adolescence, he will have to find his way through climate grief. These blue feelings will become the new normal. Mothers might feel this grief and the connected responsibilities in a profound way. To cope, and chase that elusive glimmer of hope, I find myself turning to alternative economic and political systems that centre care and community connections.

———— ⁓⋀⁓ ———— **CODE GREEN** ———— ⁓⋀⁓ ————

Circular Economies of Care

IN VICTORIA, WHERE OUR FAMILY LIVES, there is a labour shortage of childcare providers, and so finding accessible, affordable, and reliable care is a struggle for many families. In 2020, there was a shortage of 4,200 spots and a 37.3 percent access rate.[1] According to City of Victoria's head of strategic operations, Alison James, "We heard from parents 'we don't get a choice of what type of child care we want, it's whatever is available.'"[2] This was certainly our family's experience during my transition back into the workplace.

Despite putting my son's name on multiple waiting lists as soon as I became aware of his conception, I didn't manage to get access to a care facility until nearly a year after his birth. Daycare is scarce and underresourced, and until the federal government's $10/day childcare subsidy came into effect in 2022, it was expensive. Such experiences signal the pressing need for more investment into care-based social infrastructure, from daycare facilities to primary care.

This personal scenario has policy implications for many families and raises significant political questions. What would it look like, instead, as a society if we were to place a higher value on care? It would mean radically revisiting where we allocate resources and how. This means government-funded subsidies for care centres, providers, and staff, so families can afford to live and work. This would also be a way to get more women into the labour force and to bolster economies, a win-win situation for us all.

As I see it, addressing the climate emergency from a critical ecofeminist lens means emphasizing three core principles: care, justice, and community. Generally, Code Green signals a need for an evacuation. Evacuating from extractivist relations is required for centring care. This is both a profoundly transformative political undertaking and a deeply personal one.

Building up care-based economies means taking seriously social reproduction labour, including the expertise of care providers and leaders of care-based non-profits and all people, including children, women, femme, and nonbinary people, who are most directly affected by decisions about the governance of care work. As Angele Alook and co-authors elaborate in *The End of This World: Climate Justice in So-Called Canada,* with their thoughtful articulation of how to build a caring economy for all, Indigenous communities have well-developed mixed economies with care at the centre, instead of extraction.[3] Placing care at the centre of our economic systems is a radically transformative proposition. "One of the key components of climate justice," the authors explain, "is the intergenerational responsibilities for caring for future generations to live on this planet."[4] Building such vital infrastructure means investing in childcare, education, and health care and divesting from linear, extractive capitalism. This mixed economic approach, informed by Indigenous worldviews, centres the building of life-generating, reciprocal, and respectful ways of relating to lands and environments. Core to this economic system is a shift away from exploiting that which we depend on to live.

Capitalism survives and expands by treating natural resources as commodities for profit-based consumption. This transactional approach to our living environments is not the only way to organize economic systems of production. As Indigenous, feminist, and decolonial thinkers all remind us, we can collectively conceptualize care as vital political work.[5] The political theorist Nancy Fraser explains that the problem with our capitalist society is the "subjugation of reproduction to production," which undermines the affective and material labour often performed by women without pay. Fraser posits, "No society that systematically undermines social reproduction can endure for long."[6] Instead, the conditions for a deep, radical, and structural transformation of our social, political, economic, and ecological systems are needed. It's easier said than done, but one place to start is investing in care-based economies. A caring economy includes fair compensation for those providing care work, including care for children and elderly peoples, as well as investing in public spaces.[7] Investing in caring economies would mean reallocating resources, taking them away from the fossil fuel industry, the military, and prisons. Addressing the systemic issue of a patriarchal society means creating pathways toward feminist futures that build from a foundation of care, connection, and community. A caring economy entails transforming our social, political, and economic lives, moving them away from transactional, consumeristic relations.

Critical ecofeminism and the circular economy movement — in theory and in practice — offer some generative guidance on how to take small, actionable steps toward more sustainable economic relations and systems by reducing the production and consumption of goods. The circular economy movement subverts the linear economic model that sees natural resources as commodities to buy and sell — and instead focuses on reusing, refurbishing, and repurposing materials. Resisting fast fashion and carefully considering supply chains and the materials we adorn our bodies with is an issue further taken up by sustainable fashion advocates in pursuit of more circular approaches, which remind us that "sustainability is a spectrum."[8]

Many of the principles behind the notion of a circular economy resonate with kanaka 'ōiwii (Native Hawaiian) approaches to aloha 'āina — a love (aloha) for the land ('āina), or "that which feeds." Contrary to what tourism marketing materials tout, aloha does not mean an open invitation to visit Hawaiian lands as an uninvited guest. It is a term with political significance. In the words of Dr. Kamana Beamer, aloha 'āina conveys principles of respect, reverence, and justice that are vital to the healthy, flourishing, and sustainable futures of the following generations.[9] These ethics are not about a return to the past or to some imagined pre-capitalist society; rather, they are about innovation, imagining, and enacting alternative economic futures. These principles are both transformative and imaginative, offering pathways beyond the extractivist status quo.

In the early days of parenting, I wondered how I could incorporate principles of circular economies into my everyday life as a mother, while trying not to become a smug eco-conscious citizen and green consumer. Often the most organic, chemical-free, non-toxic products are beyond reach for many new parents. In small ways, I tried to incorporate circular values into my lifestyle. We took gently used spit up cloths from friends and made baby wipes. I purchased mostly second-hand items. Our car seat and stroller and some big-ticket items were previously loved hand-me-downs or purchased second-hand. When my son outgrew items, we passed them on to a friend in need or to a children's consignment store.

The root of economics draws from the Greek word *oikos* — a word that refers to both ecology and economics, pertaining to the family, property, and home. This term encompasses how we manage and live in our homes, and its reconceptualization in a more caring light can shape the futures and worlds we want to live in now and create for our children. As the COVID-19 pandemic revealed, women often take up extra responsibilities for the management of (often unpaid) household duties: educating children when schools were closed, feeding and cleaning up after members of the household and often all while continuing to work or generate income to support the family unit. Our economy would not function without this often-hidden labour. Fraser calls it a "crisis of care" in our society. Becoming a mother during a Code Red climate emergency reveals these multidimensional crises: of care for new life, both human and more-than-human, as we grapple with the future health of children and planetary well-being and a crisis of care, the climate emergency, and a global pandemic, all entangled. As such, care is a personal practice and a radical political idea.[10] And the boundaries between these realms are collapsing.

In most contemporary Western societies, care is considered cheap labour. Many nations import migrant workers to fill a care gap.[11] This is especially true for live-in caregivers and tempory workers across Canada. As Fraser says, without the affective labour of care — the labour associated with emotions, feelings, and care for others — there would be "no culture, no economy, no political organization."[12] The social reproduction of care is expansive, to include birth, raising children, maintaining households, building communities, and caring for the elderly.[13] Following Fraser, each of these forms of affective labour entails sustaining shared meanings, "affective dispositions," and associated horizons of value to underpin social collaboration and cooperation. In most capitalist societies, this labour takes place outside of the market, in homes, neighbourhoods, and civil society organizations. Struggles for the social reproduction of care work includes not just the physical act of birth, but also relates to housing, health care, food security, a living wage, migrant worker rights, daycare, elder care, and adequate paid maternity and parental leave. Ignoring the vital, affective labour of care work, which is an essential form of social reproduction, is detrimental to our survival as a society. Investing in stable care infrastructure is essential for affective labour — often relegated to women and those who govern the domestic sphere — to not be subjugated in our society.

The global circular economy movement is a worldwide movement that responds to the climate emergency through a challenge to linear, extractive systems. While ethics and principles of aloha ʻāina span generations, and are unique to the Hawaiian archipelago, there is much to learn from these movements about creating more caring, nurturing, environmentally just societies in a global context. Both consider nature to be a teacher and a source of inspiration, not a passive force or resource to be extracted. Aloha ʻāina principles and practices, like circular economies, offer lenses well situated to envision transformative and sustainable global economies. These worldviews have the potential to shift narratives and enhance conversations about the climate emergency while raising awareness and identifying alternatives. To be clear, this does not mean overemphasizing individual responsibility for environmental management, reinforcing a lifestyle blame model of environmental citizenship but, rather, thinking collectively and in community about how to reframe the ways in which humans think about and relate to more-than-human environments.

Aloha ʻāina goes deeper and further to ground a movement for ecological well-being in community, place, and sovereignty. This involves an understanding that self-determination and recognition of Hawaiʻi as a sovereign nation is rooted in the rich and vibrant lands and waters that have shaped the ecological well-being of the islands for generations. Western societies have much to learn from aloha ʻāina ethics, perspectives, and practices for the survival and flourishing of life on the planet, for planetary health and well-being.[14] While sustainability has become somewhat of a buzzword, easily commodified and marketed through "green" products, sustaining love and care for the ʻāina (land) requires a realistically radical movement for social and environmental justice. Many leaders are rising up while speaking up and out about their love for the ʻāina and wai (water), and the vitality of these life forces for the well-being of future generations. This lifespan goes much longer than that of the linear capitalist model of extraction, which takes from the environment to produce a resource and then dispose of it.

Instead of a transactional approach, we can look at our environments as life forces, rather than simply as resources for profit. This is a theme that Rachel yacaaʔał George and I take up in our reflection on and writing about fluid decolonial futures — where we discuss the animacy and agency of seascapes — and together elaborate on the importance of considering oceans as animate and full of abundant life rather than simply as highways for tanker traffic, or as empty spaces for militarization, or as resources for exploration, profit, and exploitation.[15] Examples traverse the Pacific Ocean, across various coastal archipelagoes. In Hawaiʻi for instance, the loʻi (commonly understood as a taro farm) and fishponds are examples of sustainable community development initiatives, as is the ahupuaʻa ecologically oriented governance system — a Hawaiian term that refers to a large socioeconomic, ecologic, and climatic land boundary, in turn known as palena or place boundaries or a protected place. The palena is a boundary specific to the ʻŌiwi (Native Hawaiian) system of land tenure. Kamanamaikalani Beamer explains that "These types of boundaries were not impenetrable fences, as people had the ability to move from one place to another. Instead, palena bound the material and spiritual resources of a place for community access and control while enabling chiefly leadership to regulate resources on the larger scale of the island."[16]

For an island nation dependent on local marine and terrestrial resources for survival, palena were critical for establishing manageable units for generations to thrive and survive in what might be translated today as a sustainable form of ecological governance. Mountains and waterways shape jurisdiction and governance boundaries for palena in the ahupuaʻa. An enriched understanding of the ahupuaʻa ecological system of governance requires an intimate appreciation of place.

This ecological governance approach involves a caring, embodied relationship with one's environment. This contrasts with a reality where human consumption compels an extractivist drive to take from the environment rather than to live with and learn from it. Davianna Pōmaikaʻi McGregor, professor of ethnic studies and previous director of the Center for Oral History at the University of Hawaii, Mānoa, explains: "The thing about the ahupuaʻa that is important to understand is that water is the organising principle" — jurisdiction literally flows from the mountains (mauka) to oceans (makai) and waterways to orient specific caretaking responsibilities.[17] The word for water in Hawaiian is wai, and the word for wealth is waiwai.[18] An abundance of water meant an abundance of food. These ecological boundaries run mauka (from the mountains) to makai (to the ocean) to order communities and their resources according to these land and ocean boundaries for community governance.

Capitalist development relies on a model of "take, make, use, waste." Circular economies aim to reduce waste, and aloha ʻāina practices, instead, cultivate thoughtful approaches to care in order to consider the value of natural resources throughout their life cycles.[19] As Beamer further explains, the capitalist economic system is one premised on colonialism and insatiable growth, which has destroyed ancestral systems around the world. He argues that an aloha ʻāina–based circular economy would be revolutionary. This is not some vague idea, but one with concrete policy application. Such a "give, take, regenerate" approach to a circular aloha ʻāina economic system would mean a livable wage for all, affordable housing, and universal health care. He explains how "global climate change is enhanced because modern societies have lost an intimate connection to place and the resources that feed them," thus signalling the need for aloha ʻāina.[20] Living with more caring and intimate relationships to more-than-human environments, beyond extraction and profit, to centre abundance and the health of environments is core to this aloha ʻāina way of place-based living, being, and governing.

This ecological governance approach has practical relevance for policymaking in Hawaiʻi and elsewhere. To flesh out these policy solutions to the climate emergency further,

Beamer and a group of likeminded academic-activist scholars and community leaders developed the 'Āina Aloha Economic Futures (AAEF) effort in the wake of the COVID-19 global pandemic.[21] A key concept in the AAEF Policy Playbook is that of 'āina abundance, with abundance being a concept widely emphasized across Coast Salish territory to Hawai'i.[22] In the context of Hawai'i, one goal is moving away from over-reliance on imported food — and instead cultivating locally produced food systems.[23] The forward-thinking playbook also centres Indigenous sovereignty as a foundational principle, discusses the vital significance of the Land Back movement, and returning lands occupied by the US military to the Native Hawaiian community. This emphasis is central to transformative, caring economic systems, as further elaborated by Alook and colleagues, who explain a similar vision of going beyond the capitalist, colonial, patriarchal status quo:

> We need to centre our economies on the lives of future generations, taking from the lessons of our elders. We need to realize the earth is our mother as she gives us life. We need to understand water is life and keep oil in the ground so it doesn't pollute the water. We need to show humility and respect for all our human and non-human relations.[24]

Another example of a transformative economic system is evident in some coastal Indigenous communities who embrace the potlatch system, where wealth is not marked by what one accumulates, but by what one gives away and redistributes.

The act of moving forward to contribute to raising awareness about alternative sustainable economic futures involves transformative place-based education. A pivotal starting point for learning and teaching about circular economies is in the classroom. Finland, for example, incorporates learning about circular economies at an early stage of childhood development by co-designing circular economy curricula, engaging in experiential learning activities, and taking children on tours of zero waste facilities.[25] In Finland, circular economies are associated with innovation and job creation. Companies implementing this lens are on the rise. Such holistic approaches can be successful and marketable and can enforce new ways of modifying the dominant extractivist capitalist system. In my experience, taking students outside the classroom and into community settings both in Hawai'i and Coast Salish territory to participate in service days cultivating land, removing invasive species to create space for native plants to flourish, has helped shift perspectives about how humans relate to more-than-human environments in more caring ways.

There are numerous challenges to an individualistic, extractivist, capitalist model of society. The Green New Deal transitions away from fossil fuel reliance in the United States.[26] The Red Deal movement, articulated by the Red Nation in three parts, aims to end colonial occupation while also aiming to simultaneously heal bodies and the planet.[27] Environmental justice draws into focus and problematizes an uneven distribution of sites of contamination across societies and sheds light on these inequities to create space for affected communities' visions for decolonial, sustainable alternatives. Critical ecofeminist thought challenges problematic dualisms of human mastery over nature. These critical, creative, and imaginative community-focused examples show us how circular and caring economies can be possible.

Many feminist academics, activists, and artists are speaking up and addressing the climate emergency. Women's Earth & Climate Action Network (WECAN) presented policy recommendations during the COP28 climate summit (2023), held in Dubai, based on the principles of feminist and beyond-growth economics, Indigenous rights and sovereignty, and an equitable, fast fossil fuel phase-out, among other priorities.[28] These actions complement broader global ecofeminist organizing efforts in the international arena over the years. In response to the COP26 climate summit (2021), held in Glasgow, the Policy Working Group of the Feminist Green New Deal Coalition — a group of frontline leaders, activists, and feminists who began to organize in 2019 — highlighted that the conference did not meet the urgent needs of frontline communities, and advocated for a more intersectional approach, paired with transformative leadership and an anti-oppressive social justice agenda.[29]

This coalition advocates for a radical shift from the privatization and commodification of resources to centre regenerative, sustainable, cooperative, and collective models of ecological well-being. In their words, the COP26 conference was a "step backwards" with a focus on "net-zero" and "nature-based" solutions, and an overemphasis on corporate-friendly solutions like climate financing, bolstering the problem of "carbon colonialism," with the potential to impact communities and nations in the Global South particularly unevenly.

Rather than centring environmental justice frameworks, which would emphasize human rights and returning land to Indigenous communities, COP26 emphasized and reinforced principles of extractivist capitalism. Their reflections highlighted how "billionaires, corporations and wealthy nations would be able to buy their way out of the climate crisis while frontline communities around the world and Global South nations will face its worse harms."[30] Rectifying this would involve including the policy solutions advanced by grassroots communities and gender justice advocates. As the Policy Working Group of the Feminist Green New Deal Coalition articulates, along with WECAN, feminist and environmentally just solutions to the climate emergency must be rooted in care, collective action, and community-led solutions. During COP28, WECAN organized a panel on "Tackling the Maternal and Reproductive Health Harms of Fossil Fuels and Petrochemicals," noting concern with

> economic systems tied to fossil fuel dependence [that] drive increased militarism and a mounting toll on women and people's reproductive health. A fair and fast phase-out of fossil fuels and related petrochemical and plastics (derived from fossil fuels) to a less extractive and exploitative economic model will reduce exposure and support maternal health and reproductive justice.[31]

Examples of post-extractive futures exist. The challenge of how we can bridge rich community knowledge with elite circles of policymaking and influence remains. In *Refracted Economies: Diamond Mining and Social Reproduction in the North*, Rebecca Jane Hall discusses the practices of care in post-extractive worlds based on the lived experiences of Northern Indigenous women, which emphasize community norms, relationships, and commitments where their visions for future mixed economies would be "made possible through ongoing transmission of land-based ontologies, practices and knowledges."[32] Kin networks, traditional teachings, local knowledges of plants, medicines, and beading all constitute alternative forms of post-extraction relationality and social reproduction. Care, connections, and relationships — with communities and more-than-human environments — are vital to these perspectives. Organizing efforts in community and globally, like those by WECAN and others, signal pivotal avenues for transformative economic change.

The Feminist Green New Deal Coalition Policy Working Group and WECAN organizers alike underscore how turning to climate justice solutions to address this urgent crisis requires a shift to renewable energy and a divestment from fossil fuel entanglements. This means divesting from mining, fossil fuel extraction, and an agribusiness-based economy, all of which fuel climate change. When we apply the lens of environmental justice to this issue, as scholars like Ingrid Waldron underscore through writing and visual formats, what comes into view are the ways in which racialized communities of colour, such as Indigenous, Black, and Brown peoples, are adversely affected by the oil and gas extraction sector.[33] The interactive EJAtlas (Global Atlas of Environmental Justice) documents eighty-five examples (and counting) of such sites across Canada.[34] While rich nations have committed to limiting the production of coal, they are much further behind when it comes to the oil and gas industries.

Substantive, transformative deliberations about issues related to the environmental effects of militarization require further investigation, as examples across Canada, the United States, the Middle East, and Russian reveal. In the Feminist Green New Deal Coalition's poignant words: "This posture reflects amnesia for the historic responsibility of the United States in creating climate disasters globally through war and extraction policies." Instead of this extractive approach to human-environment relations, contributors to their policy reflections summary document advocate for a "low-carbon care economy." This is one alternative approach to addressing the twin crises of care and climate.

Moments of crisis, be they in response to a sudden or slow emergency, present pivotal opportunities for critique and change. In a similar vein, with the early stages of the COVID-19 pandemic in the foreground, on April 14, 2020, the Hawai'i State Commission on the Status of Women published *Building Bridges, Not Walking on Backs,* a feminist economic recovery plan.[35] This plan fleshes out a feminist economic action plan that centres a care-based economy, which they refer to as a gender and socially responsive plan, rather than an economic plan based on extractivist principles. In this respect, the layered emergencies we face — the global pandemic and the climate emergency alike — present powerful and compelling pivotal portals to alternative futures.

The alternative care-based future advocated for by feminist-oriented thinkers and activists fundamentally challenges extractivist ways of thinking, being, and living in this world. Enhancing economic principles founded upon relationships based on care radically contend with practices of taking and extracting. Care is a radical practice in the home and in policy contexts. It is a way of being that radiates from our fluid, loving, porous bodies outward to communities across local, state, and international borders.

Critical ecofeminist thought offers alternative ways to consider how our messy, leaking bodies are fertile sites for critical reflection about our motherhood and our political lives. While, historically, Western political thought separated the unruly bodily realm from the intellectual matters of public political life, environmentally oriented critical feminist scholars challenge these dualisms. Feminist environmental political theorists Teena Gabrielson and Katelyn Parady suggest, "Traditionally, conceptions of the modern subject grant full moral agency only to those who exercise rational control over the appetites and passions of the body."[36] Instead, critical ecofeminists emphasize the vital fluidity of bodies and of their political lives.

Motherhood reveals the ways in which bodies are inherently — not instrumentally — connected to social and environmental lives. Our bodies are made of water; they carry, give, and sustain life. While classical political thought, including Western "green theory," encourages citizens to be responsible stewards as environmental caretakers — such as a strong emphasis on individual practices of recycling and reducing consumption — ecofeminist political thought provides critical reflection on this approach while challenging these binaries and underscoring the ways in which reproductive lives are always in relation to environments, embodied by, not separate from, them.[37] Instead of emphasizing the agency of rational individualism, critical ecofeminist scholarship places embodied experience at the centre of analysis. From experience comes understanding.

Vital, material, fleshy bodies are inherently connected to nature. Neither bodies nor environments are passive entities. Bodies are co-constituted with environments in a "dance of agency."[38] The nursing infant brings this into focus. The watery womb, its first environment, sustains life, feeding the fetus with essential nutrients. This dance continues through nursing in a reciprocal treaty-like, symbiotic relationship, involving a continuous connection and give and take relations, as infants communicate their needs back to their mother. Through the exchange of breastmilk, the breasts learn from the child how they are feeling, whether they are sick, need more antibodies, water, fat, or melatonin, for example.[39] Through this corporeal communication, nursing mothers transmit essential food, bacteria, and other microbes to their infants. As such, the human body is porous, resistant, "plural and connected."[40] Through the human womb and non-human liquid environment, mother and baby are united. These overlapping, messy, friendly, connected bodies show how ecosystems are part of and central to, not simply external to, our bodies. Such an emphasis on connectivity and mutually reinforcing care is a central tenet of critical ecofeminist thought. Gabrielson and Parady argue for a "fleshier democracy" as a starting point for challenging individualistic ideas and practices of environmental citizenship — noting that our bodies are embedded within environments and constitute a form of "corporeal citizenship," where our full lives are bodily, porous, plural, and inherently connected to social and natural contexts.[41] A life-giving body is a vital place to build on this fluid and emergent notion of political life. This is not to say that only mothers can be critical ecofeminists. All those with a sincere concern about the vitality and reproduction of future generations and care for the co-constitutive properties of human-environment relations make stellar critical ecofeminists.

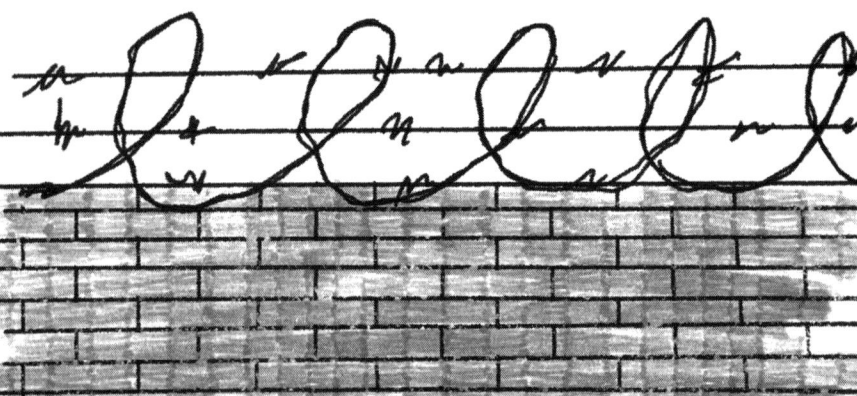

───/\\───— CODE BLACK ——───/\\───

Systemic Threats, Revealing Violences Slow and Spectacular

At 8:07 a.m. on Saturday, January 13, 2018, a text message popped up on my phone from the Hawai'i Emergency Management Agency:

> Emergency Alert
> BALLISTIC MISSILE THREAT INBOUND TO
> HAWAII. SEEK IMMEDIATE SHELTER.
> THIS IS NOT A DRILL

Immediately, my phone started buzzing with frantic calls from friends and neighbours. A new faculty member, neighbour, and friend called a few minutes after the alert went out: "Sarah, what are you doing? What should we do?" I looked out the window and saw neighbours from faculty housing spilling out of their homes in a panic. One neighbour joined another in their bathtub, and I learned later that a colleague did the same, crowding into a bathtub with his partner and children to shelter away from windows, which they had duct-taped shut. I responded to my friend slowly: "Well, right now I am in the bathroom with my phone and laptop, trying to get more information. I'm checking Twitter. Is this real? It doesn't feel real." Together, we fretted for a few minutes about what supplies to bring into our bathrooms. "I'm going to call my family," I told her, then we agreed to check in soon, and hung up the call. Like many new faculty members at the university, this was our first time living overseas, and I had a hard time processing the event. My alien, non-resident body felt detached from my surroundings.

Most of my family lives in the Greater Vancouver area, on Pacific Standard Time, three hours ahead of Hawai'i. I called my mother first. "Hi Mom. Have you seen the news? I don't want to alarm you, but we received a really distressing alert message about a ballistic missile threat heading to Hawai'i. Just wanted to call and say hi while I still can." Predictably, she started to panic. I could hear her holding back tears. "I am going to call around to different family members, Mom. I'll call you back soon." Sitting on my bathroom floor, I called my brother next, and then cycled through the rest of my family, in between refreshing my Twitter feed.

What country had I moved to? I wondered. At the time, I had only been in Hawai'i for one semester and was incredibly homesick. Never in my life had the military presence felt so all encompassing. As a privileged upper-middle-class Canadian, I wasn't used to having to fear for my life on a daily basis, but Hawai'i's geopolitical location makes it a strategic military target in the Pacific Ocean, so many Hawaiians do live with this fear. In *Detours: A Decolonial Guide to Hawai'i,* editors Hōkulani K. Aikau and Vernadette Vicuña Gonzalez explain that — in contrast to the common vision of Hawai'i as an exotic place, easy for consumption with happy hosts — in reality, "People here struggle with the problems brought about by colonialism, military occupation, tourism, food insecurity, high costs of living, and the effects of a changing climate."[1] Given rising tensions between the Trump administration and North Korea, when the emergency alert came through on that alarming Saturday morning, my colleague — a specialist in international relations — treated this alert as a very real event. For him, such an event was an eventuality.

At 8:45 a.m., a follow-up message popped up on my phone:

> Emergency Alert
> There is no missile threat or danger to the State of Hawaii. Repeat. False Alarm.

It was a false alarm, but this event showed us just how vulnerable we were as residents living in an occupied Hawai'i. It inspired poetry and innovative visual documentary storytelling to convey its significance. Kanaka Maoli scholar Heoli Osorio's poetry drew attention to this alarming scenario in a virtual reality film *On the Morning You Wake (To the End of the World).*[2] The visual documentary cites Kauai resident Cynthia Lazaroff, who stated: "The alert was false, but the nuclear threat is real."[3] Their words remind us how Hawaiian life continuously requires living on alert as a result of its heavily militarized location and a long legacy of American imperial encroachment on Kanaka ea or sovereignty, which includes treason, an overthrow of the Hawaiian political governing regime, illegal occupation, and annexation. After the Spanish-American War in 1898, Hawai'i's strategic position in the Pacific became essential to American imperial interests, and the US Congress passed a Joint Resolution that annexed Hawai'i to the United States, without consent or a fair democratic process, further shaping the military's illegitimate pathway to an occupied Hawai'i that persists to the present day.[4] Native Hawaiians became increasingly marginalized from political power. Yet, their leadership, values, teachings, protocols, and voices resound.

This is not the only possible political scenario for the Hawaiian archipelago. Local leaders and organizers offer alternative decolonial visions, visions where Indigenous people are sovereign, and deoccupation occurs through "demilitarization and the repatriation of Kanaka land."[5] At the same time, poets and community leaders remind us about possible pathways for alternative, decolonial futures. The effects of America's colonial, militaristic legacy are felt, to this day, in multiple forms given Hawai'i's strategic location for US imperialism. The military presence in Hawai'i is alarming and pervasive. It affects relationships between locals and their environments on a daily basis. Examples of this enduring presence include the leakage of Navy jet fuel from its Red Hill storage facilities into the underground aquifer affecting drinking water on O'ahu and the use of sonar affecting marine life; consequently, humans and more-than-human lives are all entangled with the potentially harmful effects of militarization.[6] The O'ahu Water Protectors, a Hawaiian activist group, took their message to the White House in 2022, demanding the investigation of the American military presence in Hawai'i, now evidently posionous beyond a metaphorical sense of the term. "Ola i ka wai!" (Water Is life) was the message they carried across the Pacific Ocean and country.[7] Incidents like the leakage of jet fuel into Hawaiian soil and waterways prompt reflection on how the American government treats Hawai'i as a parking lot for the American military:[8] it is an affront on Hawaiian lands and waterways. In Hawaiian scholar Dr. Kamana Beamer's words:

> Anyone that lives here on Hawaii Island, especially in Waimea, in my hometown, feels their home shake when they bomb Pōhakuloa. It's egregious. It hurts you deep in your soul, you know, as a Kānaka, as an Indigenous person to these islands, to feel our aina tremble and shake because of bombing and weapons of war and destruction.
>
> This issue here on Red Hill is a serious threat to the future lives and any economy on Oahu and for all of Hawaii. This is the most critical threat that we've ever had to our groundwater resources. And in fact, this is a kind of terrorist act and threat the military is supposed to protect Hawaii from, and yet they're the cause of it here on our islands. It's terrible, it's egregious, and it must stop. We must shut down the tanks immediately.[9]

In the early days of the Red Hill leak, the Navy opened fire hydrants and flushed the contaminated water right onto the sidewalks, into storm drains, and on to the ocean.[10] During the American Thanksgiving weekend, residents reported the odour of gasoline in their sinks and bathtubs. Thousands of residents were poisoned by the Red Hill fuel storage facility; however, over 100 milllion gallons of fuel remained in the tanks, posing an everyday threat to the safety of O'ahu's drinking water supply and "life as we know it."[11] The Navy committed to cleaning the mess up by July 2024, almost three years after the original incident.

Health concerns related to the militarization of lands, waters, and bodies are endless. Another contested environmental health concern is the link between ailments such as cancer in relation to military burn pits.[12] The health implications, ranging from asthma to cancer, and after-effects of exposure to toxic burn pits in war zones — "war dust" — is as much a matter of life and death as it is controversial. Toxic exposure, whether due to plastics and metals burned in war zones or effluent in places like Canada's Chemical Valley, point to the need for enhanced, not stifled, environmental health and justice research agendas.

The list of multifaceted, intersecting, structural, systemic violences continue — along with inadequate institutional responses. These converge and shapeshift in multiple violent forms: racism generally, environmental racism specifically, sexism, ecofascism. As Ingrid Waldron explains, racialized communities must confront a long legacy of state- and industry-driven approaches that consider certain bodies to be "inferior, lacking in value, and therefore expendable and disposable."[13] Acknowledging and naming, as well as addressing the systemic and root causes of environmental racism is a requirement for more environmentally just societies to thrive. Relentless risk exposure to multiple forms of violence can be debilitating and numbing. I became even more conscious of how violence manifests through various institutions and policies while living across the Pacific Ocean.

During my early days as a new faculty member, I did active shooter simulations and training, mandated after members of the white supremacist group Proud Boys threatened several instructors earlier that month. I watched in astonishment as my colleagues shot at each other with nerf guns and we took cover behind chairs and under desks in the department's main seminar room. This training took place just a few months after several of my colleagues and I gathered at a local gym to practise self-defence training to help us better prepare for direct, physical threats in our department. During my first faculty meeting in August 2017, the agenda read: "Item #1: Welcome New Faculty. Item #2: Address Neofascism." This was light years away from the portrayal of Hawai'i as a paradise. I felt very far from home and had a hard time imagining raising a family in this highly charged cultural context of violence.

While shocking when in a comparative context, we are not immune to violence in Canada. Violence takes on multifaceted shapeshifting forms — from colonialism, environmental racism, to gender-based forms of violence. The structural forms of violence in Canada occur in multifaceted dimensions as well. Take, for example, the impacts of diamond mining on lands, waters, women and culture in Northern Canada.

This violence is not natural or inevitable, but as Rebecca Hall's research reveals, it is rooted in capitalist extraction and settler colonialism. Canadians are not immune to shapeshifting forms of violence, be they slow or spectacular.

According to the Pew Research Centre, more Americans died of gun-related injuries in 2021 than during any other year on record, with a reported 48,830 deaths due to gun-related injuries. Over half of those were due to suicide. More than twelve American cities set homicide records. In some hard-hit communities, gun violence is a feature of everyday life, with residents reporting that they hear gunshots every day, and when they hear them, they know to switch locations. The lack of public visibility of these uneven relationships — where the frequency of violence is underreported in the mainstream media — has racialized roots. As the Mapping Police Violence initiative reveals, Black people are nearly three times more likely to be killed by police than white people in the United States.[14] Canadian statistics of police brutality reveal considerably high rates of violence as well.[15] Affluent white communities are less affected by gun violence, yet are more likely to influence news outlets as well as what politicians cover. Gun violence is violence fast and slow; fast for its fatal outcomes, slow for the enduring phenomenon that all too often adversely affects marginalized communities of colour.

As scholar Debra Thompson argues in *The Long Road Home*, racism is pervasive across borders, jurisdictions, and everyday life.[16] Canada is not an exception; here, violence runs deep too, and institutionalized racism has deep roots in the very architecture of the state and its administrative appendages. This is evidenced in institutional design, policy decisions, and practices over time, with enduring implications for the present. Canada's exploitation of Chinese workers as anonymous expendable labour on the railways, or the incarceration of Japanese peoples in internment camps in both Canada and the United States during the Second World War were foundational in the shaping of the state. Police violence also extends to the lived experiences of many Indigenous and Black communities. Both the Environmental Noxiousness, Racial Inequities & Community Health (ENRICH) project — a community project investigating environmental racism in Mi'kmaw and African Nova Scotian communities — and Waldron's scholarship reveal that environmental violence is a common thread running through our current climate emergency and crises-laden conditions, with racialized communities feeling the impacts of this most acutely but often with limited decision-making power.[17]

While I was pregnant at the turn of the new year as the pandemic continued, I watched the US presidential inauguration with mixed emotions. The January 6 insurrection at the US Capitol remind us that there continue to be members of society threatening the foundations of democratic institutions; these are some of the "ugliest moments" we have seen.[18] We saw this kind of democratic disregard again in Canada when a protest movement — the so-called freedom convoy — led by many truckers advocating for an end to COVID-19 vaccine mandates, took over downtown Ottawa, Canada's capital city, culminating in the federal invocation of the Emergencies Act after 23 days to end the blockade.[19] The cacophony of trucker convoy horns kept residents up at all hours of day and night. Such insurgencies challenge the health and safety of democratic life while simultaneously maintaining an indifferent and cavalier attitude toward the lives and well-being of those living and working nearby. What these events further reveal and remind us of is how law enforcement in practice conveys very different meanings, interpretations, and felt experiences for many racialized peoples.[20] Emergency declarations are not felt evenly across communities.[21] They are sometimes used by elite authorities as an administrative instrument to overstep democracy for an expedited policy outcome. At other times, they may signal a cry for help from another source of authority or jurisdiction, such as a municipality or First Nation government. They are not neutral declarations, but highly charged with political significance.

Rather than assuming individual responsibility for the well-being of ourselves and future generations, we should find solace and power in community and collective struggle. In *The Force of Nonviolence*, Judith Butler explains that nonviolence requires an end to anthropocentric individualism.[22] For Butler, "The reconsideration of social bonds as based on embodied forms of interdependency gives us a framework for understanding a version of social equality that does not rely on the reproduction of individualism."[23] Care for other forms of human and more-than-human lives and connecting the self with community highlight ways of relating beyond self-interest. Our bodies are dependent on, and connected to, other bodies for survival and well-being. Solidarity and activism emerge through multiple modes and tactics — direct action, demonstrations, speeches and spoken word, teaching, art, poetry, and legal challenges. Bodies are bound up in networks of relations to other humans, institutions, practices, discourses, and more-than-human elements, for better or worse.

Reimagining nonviolent political relations requires not just a shift in perspective about how humans relate to one another, but also to broader environments. Take Chile, for example, and the country's elected representatives' attempt to turn away from the extractivist approach to treating environments as "sacrifice zones" of the past — which included the treatment of their land as resources for easy extraction during the 1980s — to, instead, centre the vibrancy and animacy of the natural world by declaring a "climate and ecological emergency" as foundational to a proposed constitution.[24] This is one emergent example of numerous nonviolent and deeply community-oriented ways of living beyond the extractive status quo.

Dissent takes many shapes and forms. A healthy democracy surely makes space for radical and reasonable disagreement. As critical thinkers, many of us continue to challenge assumptions and question authority in a respectful and collegial manner. This looks different for everyone, and may emerge through mobilizing our bodies through protest, debate, and contributions to critical discourse through writing and academic activism. Yet, it is another thing entirely to enact violence against the underpinnings of democracy — its institutions, discourses, and practices — without any care or concern for others' well-being, as became the scenario with the freedom convoy.

These examples of systemic unrest and spectacular forms of violence shed light on multiple forms, expressions, and layers of violence and exclusion that persist in our contemporary political communities. Though healthy debate and respectful, agonistic (reasonable, argumentative) debate is necessary to enhance the boundaries of critical reflection and push decision-makers toward radical policy futures,[25] we live in a society of misinformation and distrust of science, evidence, and legislative processes. How, then, do we channel critique of systemic injustices into generative avenues, while making space for constructive, nonviolent dissent, to ultimately transform relations for more socially and environmentally just outcomes?

Reflecting on what it has been like with my son living outside of my body for as long as he was inside of me, I kept thinking about love, care, connection, and community. In *all about love,* bell hooks argues, "Communities sustain life — not nuclear families, or the 'couple,' and certainly not the rugged individualist. There is no better place to learn the art of loving than in community."[26] She does not use the term love tritely. According to hooks, justice and love are deeply intertwined: "There can be no love without justice," she famously said.[27] It works both ways. "Without justice there can be no love."[28] We learn about love — care, connection, community beyond the individual self — from the earliest days of childhood. Her writing informs us, along with Indigenous perspectives and the nonviolent stance fleshed out by Butler and others, how only when we recognize and enact a cultural perspective that respects our interconnected lives beyond possessive, self-maximizing individualism will we, as a society, be moving toward more loving, caring relations.

An awakening to love, care, community, and connection challenges status quo Western worldviews that, in turn, emphasize egotistical values of power and domination. This shift in perspective has resonance across intimate relations, between our bodies and households, to workplaces, multimedia representation, and political arenas. A loving society is more democratic and deliberative and less hierarchical. This relates to our bodies, homes, and political lives. hooks refers to these relations as a "love ethic," noting how the underlying values that we carry into our worlds shape how we think, speak and act.[29] Love as a social phenomenon — and not simply an individual practice — requires thinking intentionally about who we choose to work with, those we admire and respect, and by "committing to give our all to relationships; by embracing a global vision wherein we see our lives and our fate as intimately connected to those of everyone else on the planet."[30] Love is about connection, community, and care.

Our survival as a flourishing species on this planet depends on how we relate to each other and to our environments. Community is not a monolithic term. Community implies diversity of experiences, knowledges, perspectives, values, and ways of being in the world. Caring communities involve collective action based on honesty, open communication, and relationships that go beyond the veils or masks of performative relating. It is more than words. "We are all born into the world of community,"[31] hooks writes, and with this lived experience comes awareness of our first human experiences on this planet. For instance, a child is not born into this world in isolation, but through the portal of a birthing mother, an event imbued with messy, fluid, deeply affecting bodily matter. Children then become "born into a world surrounded by the possibility of communities,"[32] and entangled fields of relations, including, often, a team of medical professionals and family members. We are born into extended kin networks that exist beyond the confines of the nuclear family model. It is in this vibrant familial space, of broader, resonating, caring kin relations where we learn about the vital power of community.

Such loving, caring relations are fundamental to a radical shift away from the extractive power relations embedded within the deadly "-isms" of capitalism, neoliberalism, and paternalism. Friendship is an antidote to these extractive relations premised upon taking something to own, possess, or profit from. This selfish, possessive, individualistic model does not hold up when it comes to friendship, love, and care. In hooks' words: "Loving friendships provide us with a space to experience the joy of community in a relationship where we learn to process all our issues, to cope with differences and conflict while staying connected."[33] Love nourishes. It holds us up when we are feeling broken. Love is so much bigger than the individual. It emerges, radiates, and emanates outward, as hooks suggests: "Genuine love is the foundation of our engagement with ourselves, with family, with friends, with partners, with everyone we choose to love."[34] As a complement to joining broader social movements, we can all cultivate caring relations and building senses of community from wherever we are. This translates into the ways in which we greet each other, in the sharing of resources — books, meals, childcare. We can cultivate love in communities all around us, wherever we go. Going forward, inspired by this loving, this ethos of care, we can "make any place we go a place where we return to love."[35] Love moves and connects, supporting life beyond the individual to survive, thrive, feel alive.

―⋏⋀― **CODE GREY** ―⋀―

A Cautionary Tale of Renewable Extraction

THE BLURRY LINES OF EMERGENCY CODES mirror the blurry lines of motherhood. In my experience, being a new parent carries new weight with the role of trying to enact and envision healthier, sustainable habits, policies, and practices for future generations.[1] This comes with a responsibility to carefully consider the implications of our decisions and actions for human and more-than-human environments, and a concern with how to cultivate more caring relations. One pressing example comes to light for mothers, activists, policymakers, and legislators who grapple with questions about the kinds of infrastructures we need to build for the worlds we want to live in.

Specifically, by taking a close look at practices of extraction of rare earth metals, hailed as vital to the energy revolution — for e-bikes and electric vehicles, among other more "sustainable" products, I elaborate here why these vibrant materials require further investigation as a cautionary tale. In the name of green energy and clean technology, the extraction of these minerals from the earth are poised to be fraught and conflict laden for years to come. Some high profile ecopreneurs and environmental thinkers contend that electric vehicles are the way of the future for a healthier, more sustainable planet. While admirable, techno-solutions to reverse or slow down climate change deserve careful consideration and scrutiny.

There exists a widespread global belief that renewable, "clean energy" infrastructure in general (i.e., hydro, tidal, wind, or solar) that includes the powering of electric vehicles, will significantly lower emissions and address climate change. According to the United Nations, "accelerating the transition to clean energy is the pathway to a healthy, livable planet today and for generations to come."[2] But what are the implications of this in practice for planetary health? While electric vehicles trade-in fossil fuel reliance for hydro power, they are not always made without extractive relations. If one wishes to feel like an eco-conscious consumer — as I ride an e-bike daily and consider making the switch from gas guzzler to electric vehicle (EV) — it is important to think carefully about the elements, metals, and minerals gleaned from the earth to power the lifespan of critical parts, such as the battery itself. While electric vehicles may have a green reputation, their manufactured parts evoke shades of royal blue or hues of purple (cobalt) and grey (chromite, lithium, nickel). While a Code Grey generally refers to system failure, I treat this code as a cautionary signal about the promise of the techno-solutions, including the electric revolution, or "renewable renaissance," to solve the climate emergency.[3]

The future success of Canada's automobile industry will be tied to the EV transition.[4] Ontario and Québec are leading contenders for essential EV battery production plants. Overall, North American EV supply chains lag other nations and their global supply chain markets, as is the case with China and its overseas processing. Global shortages of cobalt and lithium will define the future of this contentious development, which is often liberally painted with the broad sweeping strokes of a green sustainability brush.

Andrew Nikiforuk argues that one of the prevalent challenges with this approach to "greening" automobiles is that, in reality, the EV revolution shifts humanity's expansive footprint "from one kind of mining to another, from fracked well-pads to open-pit mines."[5] EVs require about six times the minerals that a conventional automobile uses. Most of these minerals, ranging from cobalt to copper, lithium to nickel, graphite to manganese, shades of purple to grey, go into the production of EV batteries. And this is indeed a grey zone. The *New York Times* referred to the production of conflict cobalt in the Democratic Republic of Congo as a kind of blood diamond scenario, with gold-rush fervour, thus bringing with it considerable health and workplace safety concerns, along with the displacement of local residents to enable cobalt mining.[6] The military was called in to patrol a local mine. Protests led to fatalities. Human rights concerns continue to abound as children work in unsafe conditions to mine cobalt to support families given limited job prospects.[7] Cobalt is an essential part of EVs; it is used to reduce overheating and extend the range. This metal has come to be known as the "blood diamond of batteries" due to the perilous working conditions in Congo, which is the largest producer of cobalt in the world.[8] Residents and government elites rush to profit from these subterranean resources — from gold to copper to cobalt — in a competetive global market.

In addition, the production of lithium requires immense quantities of water. This scenario prompts concern that the mining of lithium is "no more green or clean than hydraulic fracturing or bitumen mining."[9] This "green solution" invades and scars the landscape and impacts the water table, polluting the earth and wells. These examples merely scratch the surface of the ecological harm done. When we dig a little deeper, the geopolitical elements paint an even more worrisome picture. For instance, most of these rare earth minerals are extracted from anti-democratic or less prosperous states, like China, the Democratic Republic of Congo, and Bolivia — often resulting in inadequate labour conditions, health and safety hazards, multiple stakeholder global interests with opaque accountability mechanisms for the well-being of local communities, and ecological devastation. The politics of rare earth mining are not necessarily poised to be more wholesome than the practices of petro states, although they might be framed as being part of the technological shift necessary to transition away from fossil fuels.[10] These highly extractive processes also produce huge amounts of waste at the sites of extraction and are not necessarily the clean, sustainable, and renewable resources that one might assume them to be in these "green" technological processes.

Thus, the extraction of earth metals and minerals is not a clean, simple, smooth process. As our societies move toward a drive for renewable resources and hydropowered electric vehicles, we face a discomforting paradox. Despite the attraction to renewables and electric vehicles, the infrastructure, processes, and working conditions involved in harvesting vital ingredients, such as nickel, is often "dirty, destructive and politically fraught."[11] In Goro, New Caledonia, mining began after Europeans arrived and colonized the region in 1853. Jurisdiction over the mining sector, with its current status as a French overseas territory in the Pacific Ocean off the coast of Australia, has and continues to be complex and multidimensional. Nickel is entangled in Goro's enduring colonial reality, where this landscape is "intimately linked to the exploitation of the Indigenous Kanak people" who call this place home.[12] Goro's earth, rich in nickel and cobalt, contains key ingredients in the lithium-ion batteries most commonly used for electric vehicles. The extraction of these highly desirable and useful minerals takes a lot of energy. This, in turn, leads to significant hazardous emissions. In Goro, nickel mining requires pumping sulphuric acid at high pressures and has led to multiple deleterious chemical spills. According to the *New York Times*, the process entails the following: "giant excavators, loaders and trucks running on fossil fuels scoop up the earth and trundle it away. Then the soil slurry is fed into a coal-fired facility that uses high-pressure blasts of sulphuric acid at high temperature to extract nickel and cobalt."[13] Then, toxic liquid waste endures in the form of tailing ponds, which local communities may be left with to manage, as foreign corporations extract from their lands.

The race for critical minerals globally affects Canada, too. Such highly extractive processes are poised to occur in Canada's rush for "green," lower-carbon infrastructure and are situated to envelop Northern Ontario through the region nicknamed the Ring of Fire — the site of nickel, chromite, palladium, platinum, copper, zinc, and gold extraction — and lands and waterways that were formerly the site of the De Beers Victor Diamond Mine in Treaty 9 territory.[14] The economic prosperity of this region is highly contested given potential environmental impacts and implications for Indigenous consent. As Emma McIntosh of *The Narwhal* highlights: "Depending on who you ask, mining the Ring of Fire region of Ontario's Far North could either help save the planet or propel it faster towards climate disaster."[15] Proposals to mine the carbon-rich James Bay peatlands are a hotly contested component of the energy transition, hailed by many political elites, including the Premier of Ontario, as a sustainable path forward for climate justice.[16] But lingering questions remain about the ecological and political implications of the development of green infrastructure for the mining of this so-called "clean technology" and impacts to Indigenous rights, lands, and waterways.

The peatlands constitute the world's second largest complex of wetlands, home to wildlife and carbon deposits. Disturbing this would potentially devastate the area ecologically. Due to the bog-like qualities of the land, it is also challenging to build infrastructure such as roads and transmission lines.

Developing the Ring of Fire region, thus, presents a catch-22 scenario: can the value of minerals extracted from this carbon sink for green technology outweigh potential ecological consequences? These developments are often framed in the language of bringing prosperity to underserviced regions for economic gains. But at what cost?

An extractive approach to mining these critical minerals implies taking from the land rather than engaging in reciprocal, life-giving relations, as many of the dozen Indigenous Nations across the 5,000-kilometre-long affected region in Northern Ontario's Treaty 9 territory advocate for. For animals such as caribou, wolverines, and migratory birds and for First Nations, these vibrant lowlands are life-giving "breathing lands" — not merely natural resources to be mined for profit.[17] Sexy promotional videos about the economic potential of the region featuring bikini-clad women offer further evidence of the potentially dangerous dual destruction caused by gender-based and ecological degradation, often linked with extractivist worldviews.[18] Despite court cases finding that government authorities did not properly consult affected communities, including the Attawapiskat First Nation, permits still allowed the prospecting company, Juno, to proceed.[19] As environmental law scholar Dayna Nadine Scott elaborates, this proposed mining of the Ring of Fire is premised on a dangerous logic of extractivism.[20] These concerns complicate the horizons of possibility for a just transition from fossil fuels to the renewable energy sector.

We must be cautious about the guise of clean energy. What would it look like to take up Scott's call to usher in "an actual post-extractive embrace of life-giving, caretaking economies that sustain us in mutual relation"?[21] Part of the response to this relies in shifting economic relations to care-based economies, centring Indigenous self-determination and agency, and aligning with alternative approaches such as the aloha ʻāina economic future orientation.

The extraction of rare earth minerals and metals presents a cautionary tale for economies trying to transition away from an overreliance on fossil fuels. Renewable energy fuelling the EV revolution may expedite overall energy consumption. Caution is needed so that electricity-powered objects and vehicles are not merely chimeric, performing an illusion of addressing climate change while exacerbating it.

With these cautions in mind, we are presented with a vital moment to imagine and enact transition pathways toward a multiscalar alternative: more care-centred sustainable and decolonial futures. At a macro (global, national) scale, this requires an investment in social infrastructure (hospitals, daycares, schools), adherence to Indigenous consent and self-determination, and centring care over capital. We must consider diversifying the way we conduct power relations in Canada and elsewhere and take seriously the perspectives and proposals of Indigenous Nations seeking to lead the way on cleaner, greener, alternative energy futures, such as implementing the UN Declaration on the Rights of Indigenous Peoples and implementing the corresponding BC legislation (DRIPA, or Declaration on the Rights of Indigenous Peoples Act), while creating space for a First Nations Power Authority.[22] At a meso (provincial, regional) policy level, this transition necessarily involves creating necessary transportation infrastructure, such as bikeable, walkable communities. At a micro (municipal, neighbourhood) level, it could include greater investment in compost and recycling programs. Alternative, sustainable, decolonial futures — what I would like to see for my offspring and other children as they grow into and create the worlds around them — require listening to innovative place-based suggestions, transforming and democratizing hierarchical policy relations to create space for community voices, implementing principles and practices of Indigenous self-determination, and co-creating abundant, healthy, vibrant, and flourishing livelihoods — not assuming that policy experts, scientists, or academics have all the solutions. This transition requires acknowledging diverse forms of place-based expertise. Alternative visions to extractive economies begin in communities, not the tech labs of machinist leaders celebrating the techno-sphere as the answer to our climate emergency.

The conflicts over nickel and cobalt illuminate how that which may seem to be clean, green, energy-efficient, and ecologically savvy resource management processes are laden with nuanced shades of grey. The clean, green technology sector along with the renewable energy sector includes more than the production of critical minerals for electric vehicles; it also pertains to how resources such as the sun, tides, and wind generate energy. Solutions to our overreliance on fossil fuels and non-renewable energy sector include investment in technological processes such as solar farms, tidal energy, geothermal, and wind farms. At the same time, these initiatives are not without contestation. Take, for example, concerns about the vibrations of wind farms for wildlife and local communities in Kahuku Valley on Oʻahu or geothermal developmant on Hawaiʻi Island, each of which has significant implications for wildlife, households, and culture.[23]

Similar concerns apply to the production of hydroelectric power. One example that is often hailed as a landmark decision negotiated by the Province of Québec and James Bay Cree is the 1975 James Bay and Northern Québec Agreement.[24] Although it is an important agreement that is often referred to as a "modern treaty" between the Québec government, Inuit peoples, and James Bay Cree, it had ecological consequences. Hyrdoelectric power historically flooded lands, homes, and Indigenous communities, often without their consent, so that the colonial state could flourish. This agreement allowed the government to continue to invest in the construction of hydroelectric dams while offering financial compensation to the affected Indigenous communities. Moving forward for transformative, more sustainable economic systems, diverse and circular economies are needed that create avenues for collaboration, co-governance, and respect for Indigenous leadership when it comes to power relations in both senses of the term — physically (hydroelectric) and metaphorically for the relations that shape the governance of lands, waters, and ecosystems.

Other examples draw attention to the adverse impacts of hydroelectric power on ecosystems in Canada and globally. In Albania for example, the 500-metre-wide Vjosa River, which exits at the Adriatic Sea, is home to diverse plant and animal species. It is a lifeline for multigenerational farmers, now facing the radical changes imposed on their ways of life due to the proposed development of 400 new hydroelectric power plants.[25] For many of those living along the Vjosa, the river is life. Their livelihoods are inherently connected to the river. Rivers in general are vibrant sites of life, be it animal, plant, human, or more-than-human. This vibrancy is poised for deterioration with the imposition of infrastructure like a dam. As a society, we are in a bind then: there is a need for clean energy and sustainable futures for our children, while at the same time, a need to apply caution to initiatives that impose on and constrain ecological life and well-being. This is the fundamental problem associated with treating natural resources as instrumental commodities for sale and profit in our contemporary capitalist system. We could turn in a different economic direction that directly challenges extractive capitalism. What if, instead of looking at waterways as resources, we were to think of these vital ecological entities as life-giving sources? What alternative power relations and generative futures for energy initiatives might power healthier human-environment conditions? These are the kinds of questions that, along with critical ecofeminist, ecosocialist, and environmental justice scholarship, our children will be taking up for generations to come, as Waldron's research so poignantly draws attention to.

Coded relations — red, orange, pink, blue, green, black, grey — fall along a spectrum; they intersect, and signal the need for attention to the vital, pressing, and immediate ecological matters of our times. This closing vignette blends codes together, mixes colours, and begins to make some suggestions for alternative ecological futures. While we must take seriously any State of Emergency declaration, be it issued by a municipal, Indigenous, provincial, federal, county, state, or international agency, we must also ask hard questions about what these declarations mean, how they relate to democracy, and whose voices are included and excluded in the decisions that follow. This is critical, vital even, to moving through and beyond the climate emergency — a twin crisis of climate and democracy.

As I discuss from my experience and perspective as a new mother living through a Code Red climate emergency, I also feel, observe, and witness the layered realities of multiple other coded emergency events, sometimes swiftly, sometimes slowly. By carefully reflecting on the experiences, timing, sequences, severity, implications, and management decisions related to coded emergencies, *Hot Mess* both calls attention to the intensity of the climate emergency that intersects with each of the codes discussed in this book and calls for action as we contemplate alternative ecological relations for future generations. Whether reducing our fuel consumption, engaging in circular economies, supporting Indigenous self-determination, or monitoring energy consumption, there are many ways we can individually and collectively engage in more caring relations with each other and the planet.

While the experience of a "hot mess" mother entering the early stages of motherhood parallels the blurry, overwhelming, infuriating realities of the contemporary climate emergency — each hot mess experience is not just comprised of trauma, sadness, or despair. My experience as a new parent involves so many moments of joy and beauty, even with the bleak backdrop of climate change. I learned to care in new ways while sharing my body in reciprocal life-giving relations. Perhaps if we could take these mutually beneficial relations between a parent and infant as an organizing symbol of our political relations — premised on care, rather than self-interest, profit, or extraction — we could embody, enact, and envision healthier, generative, sustaining relations into the future. Beginning with our bodies, our homes, our everyday practices, both individually and in community, we can create more abundant relationships between ourselves and planetary lives to address the collective crises we face and live well with one another.

PRISMATIC REFLECTIONS

Cultivating Care and Community through Multifaceted Crises

WHETHER WE ARE CONSCIOUS ABOUT THIS OR NOT, we feel climate change in and through physical bodies. The climate emergency is a public health emergency. Bodies hold multilayered emotions, complexities, and crises in tension. An emergency room is where this tension often manifests and a short-term, hit-or-miss remedy applied. Slow emergencies require the emergence of alternative treatments, imaginative solutions, and ongoing relationships. As a new mother, with a nursing body and without a primary care provider, I ended up in the hospital seeking emergency care. This likely wouldn't have happened if I had a meaningful and continuous relationship with a family doctor. What I learned through this, thinking about the parallels of my layered corporeal trauma, first through the emergency Caesarean then the visit to the hospital emergency ward, is how emergencies are expository techniques, shedding short-term light on the shadows of persistent pain.

Slow emergencies are power-laden. Some beings have access to treatment and care more readily than others. This is something that academics and activists from critical disabilities studies communities draw attention to.[1] This critical policy studies lens, bridging disability and climate justice, signals the need to rethink expertise and diverse forms of situated bodies of knowledge beyond elite, hierarchical, top-down ways of knowing. The perspectives, stories, and voices of those affected by multifaceted policy problems — public health crises, the climate crises, access to housing, and so on — are vital for policy justice.

No singular hospital code quite captures the experience of living through the dramatic event of a public health or climate emergency. In my experience as a pregnant then nursing mother, over the course of a few months, I lived through Code Red and Code Pink emergencies, all the while feeling then reeling from the layered effects and affectedness of multiple, intersecting, overlapping, entwined, multifaceted emergencies simultaneously — the climate emergency, a global pandemic, and a traumatic birth. The simultaneity of the emergencies invokes prismatic thinking. Far from being stand-alone crises, the emergencies we face are multidimensional and multifaceted, requiring no one angle of vision, but a prismatic lens. Reaching for and orienting to more environmentally just futures for subsequent generations necessitates a thoughtful approach to how we, as a society, respond to emergencies that are both swift, spectacular, and immediate and those that are more systemic, slow-moving, and seemingly subtle or quiet. All of these draw into focus a need to re-envision and re-enact a spectrum of generative, reciprocal, and life-giving relations between humans and more-than-humans.

Emergency health codes require further attention and nuance because we often feel multiple coded experiences in our bodies at once, whether we realize this or not. Such visceral, deeply affecting embodied experiences necessitate further consideration as we carefully look to the future and envision caring, healthy, sustainable alternatives to the status quo for future generations. A code is a cue to alert attention, direct resources, and then when the fast and intense pace of an emergency subsides, contemplate how such circumstances might be addressed or avoided altogether. The term also suggests something obscured from common view or common knowledge. Emergency codes are also cues. They draw our attention to systems and infrastructures that affect human and planetary health. In the heat of the moment, there is often not much time for contemplation or reflection, but an impetus to act quickly and swiftly to manage and mitigate any further harm. Thinking carefully about embodied experiences of the climate emergency, be it coded red, orange, pink, blue, green, black, or grey, provides a multifaceted prismatic lens into the perspectives of how we might learn from these deeply affecting corporeal circumstances to envision possible futures that are less emergency-laden, fraught, and extractive, and instead more caring, sustainable, and ultimately more just.

One code can carry many meanings. For instance, in British Columbia, Code Grey refers to system failure, like the disruption to electricity, water, communications, or information technology, but the meaning varies across the country.

In the city of Sarnia, located on traditional Aamjiwnaang territory adjacent to Canada's Chemical Valley, the local Bluewater Health authority declared a Code Grey due to a utility/infrastructure breakdown in 2014 when over 500 kilograms of hydrocarbons spilled into the air from Imperial Oil, setting off an alarm.[2] This code generally refers to a critical system failure or loss of infrastructure in a hospital setting. The plant went into lockdown and workers evacuated, but the scent of gas enveloped adjacent neighbourhoods. Residents reported getting headaches and nausea. But local air monitoring did not detect anything in the air, and an all clear was issued later that evening. Those living in Chemical Valley also have to deal with the opaque CVECO (Chemical Valley Emergency Coordinating Organization) code system, ranging from the least urgent Code 5: "Incident outside of industry potentially requiring CVECO assistance," to the most pressing Code 9: "Request for response by Municipal Fire Department."[3] Sometimes — and it's not always clear why or what to do when this occurs — sirens may sound, and residents are encouraged to shelter in place.[4] Citizens of the Aamjiwnaang First Nation constantly live through emergency conditions in their everyday experiences.

Coded red, orange or grey: these layered realities call attention to how the climate emergency can be felt differently whether one lives in Chemical Valley, where chemical release alert warnings are common, or in British Columbia, where heat waves, wildfires, and flooding of entire cities are still an exception to the norm, with limited alert messaging systems in place. The issuance of a code is both an alert mechanism, a tool of emergency management and response, and a metaphor for the coded, confusing, hazy atmosphere, tone and sensations of living through multiple aspects of emergency life with different sensory experiences and tempos. They can be simultaneously swift, requiring immediate evacuation, and slow, as the climate emergency is one that requires systemic, transformative action.

Emergencies are multidimensional and layered. Not everyone experiences an emergency the same way. As I've observed and witnessed through my research on State of Emergency declarations in communities like Attawapiskat, addressing systemic slow violence against Indigenous peoples is connected to responses to the climate emergency, as Cree environmental activist Clayton Thomas-Müller argued in 2022 in the *Globe and Mail*, urging public officials to engage seriously with a Just Transition agenda, which includes moving away from fossil fuel reliance to more renewable forms of energy generation while ensuring Indigenous self-determination.[5] This transition requires respect for Indigenous sovereignty as well as ensuring consent for extraction initiatives, such as in the Ring of Fire. Indigenous communities and perspectives about imaginative, alternative decolonial futures, expressed through op-eds, policy reports, poetry, music, and even jurisdictions, show that the world we live in, in all the multifaceted ways in which we experience life, does not have to be this way. There is great power in collective action and collective imagination. At the core of this book is an emphasis on cultivating relationships of care with each other and our living environments instead of extractive relations to envision and enact sustainable futures for upcoming generations. I write from my vantage point as a new mother and critical ecofeminist scholar concerned about the climate and committed to alternative habits, ideas, laws, policies, and practices such that we can create and live brighter, healthier, and more sustainable lives now and for the generations to follow.

Climate change is an urgent, pressing emergency that becomes evident through cascading disasters. The 2021 New York floods, which trapped residents in their basements and revealed a systemic lack of extreme weather readiness, is one poignant example.[6] In the aftermath of disasters, financial ruin follows for many people, pushing communities to the brink of insolvency.[7] Many societies are unprepared for these events, deferring critical maintenance until it is too late. Soon enough, vulnerable coastal communities will be pressed to pay for costly insurance to adapt and mitigate their risk exposures to the extreme effects of climate change.[8] Flood insurance will be a new reality for many trying to adapt to their new, wet existences. Communities in these conditions are running up against the edges of resilience. We are living in times beyond adaptation. Governance and regulation cannot keep pace with a rapidly changing climate.

In Coast Salish territory, one of the persistent emergencies I took note of throughout my pregnancy and into the early stages of my son's arrival in this world was the nation's longest act of civil disobedience at Fairy Creek, a movement against logging British Columbia's treasured old growth forest. Indigenous and non-Indigenous environmental defenders stood firm in Pacheedaht and Nuu-chah-nulth territories, holding their ground to draw attention to this vital resource, as a last resort, clinging physically and symbolically to a living harbinger of hope against the pressing realities of a rapidly changing climate. This intergenerational, nonviolent act of resistance took many forms. Protesters against the logging occupied Johnson Street Bridge in Victoria, British Columbia, painting their hands red to call for accountability for over 1,000 arrests.[9] RCMP officers broke protest lines with pepper spray.[10] Reporters were blocked off from covering the protests at Fairy Creek, sequestered to an exclusion zone.[11] This example — the protest and its suppression — draws into sharp focus what happens when citizens who care deeply about their ecological futures and the health of planetary lives feel silenced, their voices unheard by authorities and decision-makers. Democracy, transparency, and accountability are at stake in these struggles.

What does it take to enact what ecofeminist scholar and activist Vandana Shiva calls an earth democracy, to do away with the carceral approach of arresting and punishing those standing up for the threatened environments? An earth democracy, as Shiva explains, stems from the foundational principle that humans are a continuum of their environments.[12] So what might alternative ecological futures look like in practice from this vantage point? Further still, how might we envision a parliament of vital things — a speculative project that invites all things, animals, and plants into parliamentary settings — as political ecologist Bruno Latour has proposed?[13] This creative, imaginative project asks participants to consider and enact laws and governing relations beyond men, to centre life. It invites the audience to envision parliament beyond a separation between culture and nature. Latour and collaborators ask: "What if we welcome all things into our parliament? What would be the plight of the planet? The reasoning of a fish? What claims would trees make, and what future would oil see for itself?"[14] There are many ways we can reimagine and enact more caring, connected political and ecological relationships with more-than-human lives. How might we reconceptualize law like a living treaty, as Indigenous legal scholar Aimée Craft suggests, such that it can be understood as a living force, breathing life into those very environments under siege?[15] Each of these ecological thinkers encourages us to take the liveliness of our animate environments seriously.

Concern about the well-being of our forests is not a stand-alone event. This becomes clear in the actions of environmental advocates organizing tree sit-ins, high above the ground, from the tree canopies of maple and cottonwoods.[16] Land defenders remain hopeful that their embodied actions will hold off development of the expanded Trans Mountain Pipeline even further, a site poised to threaten the ecosystems and livelihoods of many communities along the oil's path as it makes its way from the Alberta oil sands to the Salish Sea to be refined for global markets. Through direct action like a sit-in or community-led video, Indigenous and non-Indigenous peoples are raising their voices about the potentially adverse effects of this extractive infrastructure and shifting consciousness about the vital voices of more-than-human environments whose well-being directly affects our own. Their interruptions of these petroleum lines simultaneously challenge any rigid demarcation between human separation from nature, an age-old binary long contested by ecofeminist, Indigenous, and environmental justice scholarship and activism. Nonviolent direct actions from Attawapiskat to Coast Salish territory, where the final stages of this writing took place, illuminate a deeply loving and caring approach to the local environment, with human and more-than-human bodies organizing at the front lines. Standing up for what matters — vital forest ecologies, a deep love for the ecosystems that feed and nourish societies — in these cases serves simultaneously as a physical and metaphorical signal about the visceral consequences of the current climate emergency.

Relationships are central to earthy tree networks of kinship and care as well as community-engaged research and public engagement. Forest ecologist Suzanne Simard explains how trees talk and communicate, illuminating the vast ways in which forests are homes for bacteria and fungi with the capacity to cycle and decompose carbon, build soil, and filter water. They are the "caregivers of the forest," and humans have much to learn from the legacies of forests for healthier human/more-than-human relations. This requires the humility of listening to and thinking carefully about how to create productive environments for trees to grow. Trees mother; they pass along wisdom in networks of kinship and reciprocity.[17] Humans have much to learn from trees about high-functioning and networked communities. Humans need to de-centre their atomistic egos and learn more from trees about symbiosis and regeneration, vitally how to live well on this planet now and for future generations. Indeed, the perils of deforestation would be hugely destructive to humans and more-than-humans. The ravaging fires that took over the Pacific Northwest during the summer of 2021 herald what is to come if we take the teachings that trees offer for granted when faced with our rapidly changing climate.

The elements — air, earth, fire, and water — draw attention to the ways in which humans are not separate from their surroundings, but reliant on and entangled within them for the sustenance of life itself. Take air for example. Pregnant and with limited mobility, in the late summer of 2020 when wildfire smoke choked residents around the Pacific Northwest, I wandered around a mall to get my steps in and breathe the cleanest air available while wearing a mask to protect myself as the global pandemic raged. During these wanderings, I contemplated the daily realities for communities around the globe living with pollution affecting their everyday lives. Most notably, as elaborated in the "Code Orange" vignette, based on my relationships and research, I thought about and continued to advocate for more meaningful, transformative responses to the pressing challenges presented by the chemical plants that encroach on the lived experiences of the Aamjiwnaang First Nation near Sarnia, Ontario, at every turn.[18] Breathing, in these conditions and elsewhere, is a political act. Communities should not have to fight for the inherent and fundamental right to breathe clean air.

Then there is earth. How the planet's early elements — lands and soils — face duress receives considerable coverage in the press relative to elements. Writing the concluding passages of this book while in my home on traditional and unceded Coast Salish territory — home to many ləkʷəŋən-speaking peoples as well as the Songhees, Esquimalt, and W̱SÁNEĆ peoples whose historical relationships with the land continue to this day — I am reminded of the abundant and generative relations that many Indigenous peoples articulate with reverence for living environments, sometimes referred to as Mother Earth. The symbol of Mother Earth also features prominently in a variety of settings and protest signs. These actions prompt critical reflection, such as what it means to be a "voice for the forest" as the Sierra Club suggests.[19] Thinking about human/forest relations differently requires a reorientation of relationships, a necessary turn to relational ontologies and cultivating spaces and practices for listening, witnessing, and changing behaviours and perspectives. This foundational, pivotal, relational approach chimes with the kind of earth democracy proposed by Shiva.

Then there's fire that leads to breath strained, lungs clenched; forests burned during the height of wildfire season in 2020 and then again, as I write on the tailwinds of a provincial State of Emergency due to wildfires in British Columbia. As I became a new mother, struggling to nurse my newborn, cities were engulfed in flames and lost. Same scenario a few years later as I was raising a toddler. My heart aches for Maui, Yellowknife, the Okanagan. Climate emergency events that once felt exceptional now mark a new normal. In 2021, heat waves ballooned above us. When the wildfires ripped across the Pacific Northwest, many Indigenous communities were ahead of the game, with years of experience conducting controlled burns.[20] The catastrophic event of extreme heat presents a critical opportunity for learning. Western societies have much to learn from Indigenous communities with long-standing practices of fire control and management.

Finally, there's water. Our bodies are made of water. When planetary systems are threatened, heavy rains pour down, flooding communities. Glaciers melt. Oceans acidify and warm. Oceanic volcanoes engulf islands. Hurricanes wreak havoc. Abandoned pipelines harm the seafloor, as we saw in the Gulf of Mexico.[21] Sea levels rise. Fishers struggle to harvest fish. When the 2021 heatwave ballooned over the Pacific Northwest, more than a billion sea creatures suffered.[22] Water shapes how we enter the world. It carries the potential to create and destroy life. We emerge into this world through the liquid environment of our mother's womb. The birthing body is a liminal threshold between worlds. We are a sea of cells. In Terry Tempest Williams' words: "In this vast, undulating ocean, we are cradled. The waves carry us like the rise and fall of the melody of mothers."[23] Much of who we are originates in water.

As I walk along the coastlines near where I live, I am reminded of how mothers and infants are transitional environments holding tensions in place, cradling worlds. Like me, my son was born on the edge of the Pacific Ocean. At this edge, an intertidal site of play and joy orients us to the horizons, to future possible worlds. Together, while in utero and throughout the early days of his life, we would walk along the shoreline, falling in love with each other and the sea. I would play, pregnant, in the waves. This moving, fluid relationship between mother, body, and sea would be foundational for our family. Following Williams: "We are born from what is fluid, not fixed. Water is essential. Mother is essential. The ocean as mother is mesmerizing in her power, a creative force that can both comfort and destroy."[24] The mother's body is a vital lifeline, a portal, a sanctuary, and a refuge.

A writer's words emerge like water. When a woman's water releases, she goes into labour and birth is generally imminent. When a writer's imagination breaks loose, so too emerges labour of another sort. In my life, water provides reprieve. By walking shorelines, I find solace and respite from the intense heat, pressures, and fires of our contemporary worlds. While I walk or jog along the ocean, alone or with my infant son, I catch my breath, pause, and contemplate how to keep my cool with the layered and seemingly insurmountable challenges presented by the climate emergency.

At the edges of Victoria's coastlines, I face international waters, look to the mountains, and wonder about what alternative ecological worlds might be possible. These moments give me life. They sustain the core of my being. I look to my son as he's swaddled and faces me in his stroller, and observe his sense of wonder as he takes in his surroundings. It is this love for one's surroundings, an appreciation for our vibrant environments, that I want him to notice. For when we notice, we feel something, and we begin to care. Throughout pregnancy and during the first year of our time together, my son and I would share many moments like this. Cultivating this kind of ecological awareness, from womb to worlds, carries the transformative potential to build connections, challenge extractive relations, and centre care. Such an ethic of care

— love for one's child, for our environments, for communities — may seem simple, but it is also a radical approach. Care is at once deeply personal and profoundly political. When we care about each other, and nurture life-giving, generative relationships, we are all better off.

Relations of care emerge in many forms. It can mean caring for a child. Care may also mean providing meals, babysitting, or gift-giving. Living and working through the entwined hot messes of parenthood and the climate emergency teaches me many lessons about decentring my ego and how individual action and responsibility alone will not save us or the world. Parenthood, friendship, and community are all relations of care that extend beyond atomistic individualism. When we extend care to each other, we engage in radically transformative relations. Whether we are tending to a community garden plot, participating in a car share program, or contributing to a circular economy and repurposing goods, we are living in community, demonstrating relations of care. To navigate beyond the worlds created by overconsumption, fossil fuel extraction, and individual accumulation of wealth, we can learn from the early days of parenthood how to better and more lovingly, joyously care for each other. Our bodies are always in relation to our environments and kinship families. Moving away from the hot mess of climate change requires centring care over consumption. The messy, loving, emergent care work is central to human existence and vital to the flourishing of our political lives as humans embedded in more-than-human worlds. To the core, vitality and prosperity require relationships premised on connection, community, and care.

Acknowledgements

Much like any labour of love, this book did not emerge in isolation, but in community. I am grateful for the enduring support, messages, phone calls, conversations, meals, hugs, and hangouts with family and friends who carried me through a pandemic pregnancy as the world burned and wildfire smoke blew into my apartment. It is a privilege to be able to cultivate a sense of home on unceded Coast Salish lands and waters along the Salish Sea that have held up Lək̓ʷəŋən and W̱SÁNEĆ Peoples, including Songhees and Esquimalt Nations, whose historical and enduring relationships continue to this day. With this privilege comes the responsibility to treat our shared environments better for future generations and to contribute to broader awareness of and education about Indigenous sovereignty and self-determination. I am fortunate to continue to learn with and alongside many Indigenous leaders and scholars who have radically reoriented my thinking about more caring relationships among humans and more-than-human worlds. Special thanks to my seascapes crew, Rachel yacaaʔał George, Dawn Smith, Robina Thomas, and Jen Bagelman for holding space, lifting me up, and paddling together during trying times. I also express gratitude to the FERN (Feminist Environmental

Research Network) community and critical ecofeminist co-conspirators, especially Jennifer Lawrence and Emily Ray, for sharing so many laughs and life lessons, all while trying to refuse extractive capitalism and to centre care for one another and the worlds around us.

As I learned in visceral detail, birth is an event that holds together complex feelings about loss and life. My traumatic birth experience fuelled these pages during a tender postnatal period as I navigated unfamiliar waters, at high risk for postpartum depression while an unprecedented heat dome ballooned in British Columbia. Thank you to my support team, the Midwives Collective, my birth doula, and my partner, John, for standing by through vital stages of gestation during the perinatal period, including life inside and outside the womb. The walks that became waddles, the swims that became swift splashes and the winter milkshakes that satisfied corporeal cravings all kept me afloat during an intensely transformative period of life change. An overseas relocation during a global pandemic, a new job, and a budding family all contributed to significant changes that inspired me to write about and reflect on how a Code Red emergency becomes an embodied event, something deeply affective, reminding us just how interconnected our bodies are to broader environments.

I am grateful to the team at Fernwood Publishing for seeing the potential in this project and for their unwavering commitment to supporting critical books for critical thinkers. Tanya Andrusieczko believed in this project from conception to creation, and her sharp editorial vision refined the pages into something accessible to diverse audiences within the academy and beyond. I sincerely appreciate this generative process of co-creation, sharing ideas and perspectives and always refining the focus of core messages that follow. Tanya's insight helped clarify that this book contributes to radical critiques about extractive capitalism so that caring worlds can flourish. I hope that by sharing these stories about encounters with colour-coded emergency experiences, more conversations about care, connection, and community will emerge so that the generations that follow do not just survive but thrive.

Endnotes

PREFACE: A STATE OF EMERGENCY AS AN EMBODIED EVENT

1   *BBC News*, "Canada Weather: Dozens Dead as Heatwave Shatters Records," June 30, 2021, https://www.bbc.com/news/world-us-canada-57654133. For further coverage of climate disaster events, see Tyee coverage and the "Bracing for Disaster Series" (https://thetyee.ca/Series/2023/05/15/Survivors-Climate-Disaster-Series/) in partnership with the Climate Disaster Project at https://climatedisasterproject.com/.

2   Raymond Zhone and Elena Shao, "2024 Begins with More Record Heat Worldwide," *New York Times,* February 7, 2024, https://www.nytimes.com/2024/02/07/climate/2024-hottest-january-data.html; Christopher Flavelle, "Wildfire Smoke Will Worsen, New Study Shows, and Protections Are Few," *New York Times.* February 12, 2024, https://www.nytimes.com/2024/02/12/climate/wildfire-smoke-health.html.

3   Penny Daflos, "Dying for Weeks: New Details Emerge as B.C. Heat Dome Death Toll Rises," *CTV News,* November 1, 2021, https://bc.ctvnews.ca/dying-for-weeks-new-details-emerge-as-b-c-heat-dome-death-toll-rises-1.5647914; BC Coroners Service, *Extreme Heat and Human Mortality: A Review of Heat-Related Deaths in B.C. in Summer 2021*, Report to the Chief Coroner of British Columbia, June 7, 2022, https://www2.gov.bc.ca/assets/gov/birth-adoption-death-marriage-and-divorce/deaths/coroners-service/death-review-panel/extreme_heat_death_review_panel_report.pdf.

4   BC Coroners Service, *Extreme Heat,* 5.

5 Michelle Gamage, "It's Official. Climate Change Has Brought Deadly Health Risks to BC," *The Tyee,* February 14, 2024, https://thetyee.ca/News/2024/02/14/Climate-Change-Deadly-Health-Risks-BC/.

6 Michael Joseph Lee, Kathleen E. McLean, Michael Kuo, Gregory R. A. Richardson, and Sarah B. Henderson, "Chronic Diseases Associated with Mortality in British Columbia, Canada, during the 2021 Western North America Extreme Heat Event," *GeoHealth* 7, 3 (March 2023), https://pubmed.ncbi.nlm.nih.gov/36938119/.

7 BC Coroners Service, *Extreme Heat,* 5.

8 Sarah Grochowski, "Heat dome to Polar Vortex: 5 Extreme Weather Terms You Should Know," *Vancouver Sun,* November 12, 2021, https://vancouversun.com/news/local-news/heat-dome-pineapple-express-water-spout-extreme-weather-events-a-new-normal-in-b-c.

9 Timothy Luke, *Anthropocene Alerts: Critical Theory of the Contemporary as Ecocritique* (Candor, NY: Telos Press, 2019).

10 Emily Oster and W. Spencer McClelland, "Why the C-Section Rate Is So High," *The Atlantic,* October 17, 2019, https://www.theatlantic.com/ideas/archive/2019/10/c-section-rate-high/600172/. See also World Health Organization, *Caesarean Section Rates Continue to Rise, Amid Growing Inequalities in Access,* June 16, 2021, https://www.who.int/news/item/16-06-2021-caesarean-section-rates-continue-to-rise-amid-growing-inequalities-in-access.

11 Zoanne Clack, "Women's Health Concerns Are Dismissed More, Studied Less," *National Geographic,* December 17, 2019, https://www.nationalgeographic.com/magazine/article/womens-health-concerns-are-dismissed-more-studied-less-feature.

12 Further details about the Code Orange declaration available online. See Arthur C. Green, "Kelowna General Hospital Issues Code Orange for 'Mass Casualty' Event," *Western Standard,* August 18, 2023, https://www.westernstandard.news/bc/breaking-kelowna-general-hospital-issues-code-orange-for-mass-casualty-event/article_ca6512e6-3e08-11ee-aba6-8f31dfca15d7.html.

13 See Premier David Eby's statement, Office of the Premier, "Premier's

and Minister's Statement on Provincial State of Emergency," https://news.gov.bc.ca/releases/2023PREM0054-001323.

14 William J. Ripple, Christopher Wolf, Jillian W. Gregg, Kelly Levin, Johan Rockström, Thomas M. Newsome, Matthew G. Betts, Saleemul Huq, Beverly E. Law, Luke Kemp, Peter Kalmus, and Timothy M. Lenton "World Scientists' Warning of a Climate Emergency," *BioScience* 72, 12 (2022): 1149–1155, https://academic.oup.com/bioscience/article/72/12/1149/6764747.

15 See, for example, Sarah Marie Wiebe, *Life against States of Emergency: Revitalizing Treaty Relations from Attawapiskat* (Vancouver: UBC Press, 2023). And see also the project story map with excerpts: https://storymaps.com/stories/3e0020d699034eb7bd56d7df816b45c0.

16 David Uahikeaikaleiʻohu Maile and Sarah Marie Wiebe, "When a State of Emergency Is Declared, We Should All Be Alarmed," *Abolition* (blog), September 27, 2019, https://abolitionjournal.org/when-a-state-of-emergency-is-declared-we-should-all-be-alarmed%EF%BB%BF/; Peter Baker, "The President's Decision Incited Instant Condemnation from Democrats, Who Called It an Unconstitutional Abuse of His Authority and Vowed to Overturn It with Republican Support," *New York Times*, February 15, 2019, https://www.nytimes.com/2019/02/15/us/politics/national-emergency-trump.html; UN Environment Program, "Facts about the Climate Emergency," https://www.unep.org/explore-topics/climate-action/facts-about-climate-emergency.

17 Michael Orsini, "Feeling Critical: Navigating the Emotional Worlds of COVID-19," *Critical Policy Studies* 15, 3 (2021): 387–397, 387.

18 Sarah Marie Wiebe, *Everyday Exposure: Indigenous Mobilization and Environmental Justice in Canada's Chemical Valley* (Vancouver: UBC Press, 2016).

19 Sarah Marie Wiebe, "Crisis Atmospheres: Sensing Life on Alert, A Visceral Response to Timothy Luke's Anthropocene Alerts," *New Political Science* 42, 4 (2020): 601–610.

20 For an elaboration of these concerns, see the *Civil Beat*'s discussion from July 2023: Kristin Downey, "Governor's Housing Proclamation Draws Skeptical Response from Oahu Officials," *Honolulu Civil Beat*,

July 19, 2023, https://www.civilbeat.org/2023/07/governors-housing-proclamation-draws-skeptical-response-from-oahu-officials/.

21 Wiebe, "Crisis Atmospheres," 4.

22 Judith Butler, *The Force of Non-Violence: An Ethico-Political Bind* (London: Verso, 2020).

23 Wiebe, "Crisis Atmospheres," 8. The political environment leading up to and following the emergency of Theresa Spence's ceremonial fast is discussed at length in my forthcoming book with UBC Press: *Life against a State of Emergency: Reverberations of Treaty Relations from Attawapiskat.*

24 United Nations, "Secretary-General Calls Latest IPCC Climate Report 'Code Red for Humanity', Stressing 'Irrefutable' Evidence of Human Influence,'" Meetings Coverage and Press Releases, August 9, 2021, https://www.un.org/press/en/2021/sgsm20847.doc.htm.

25 Raymond Zhong, "Climate Scientists Warn of a 'Global Wildfire Crisis,'" *New York Times,* February 23, 2022, https://www.nytimes.com/2022/02/23/climate/climate-change-un-wildfire-report.html. See also Ripple et al., "World Scientists' Warning."

26 Rob Nixon, *Slow Violence and the Environmentalism of the Poor* (Cambridge, MA: Harvard University Press, 2011).

27 For more about our seascapes research and collaborations, see https://seascapesstories.ca/. For an elaboration of yaʔakmis, see Dawn Marie Smith, "hiił kʷiiʔił siƛ (Bringing Something Good from Way Back): A Journey to Humanize Post-Secondary Education" (doctoral dissertation, University of British Columbia, 2018), 80. See also Richard Atleo, *Tsawalk: A Nuu-chah-nulth Worldview* (Vancouver: UBC Press, 2004).

28 For an elaboration of "takotsubo," or "broken heart syndrome," see Thomas Verny, "Understanding the Reality of Broken Heart Syndrome," *Globe and Mail,* February 9, 2024, https://www.theglobeandmail.com/life/health-and-fitness/article-understanding-the-reality-of-broken-heart-syndrome/.

29 See *Seascape Stories: Decolonial Water Relations* https://seascapesstories.ca/ and also Rachel yacaaʔał George and Sarah Marie Wiebe,

"Fluid Decolonial Futures: Water as Life, Ocean Citizenship and Seascape Relationality," *New Political Science* 42, 4 (2020): 498–520, https://doi.org/10.1080/07393148.2020.1842706

30 Ingrid Waldron, *There's Something in the Water* (Halifax & Winnipeg: Fernwood Publishing, 2018).

31 Environmental Noxiousness, Racial Inequities and Community Health (ENRICH) Project, https://www.enrichproject.org/about/background/.

32 Francesca Fionda, Jeffery Jones, and Chen Wang, "B.C.'s Multimillion-Dollar Mining Problem," *Globe and Mail,* February 21, 2024, https://www.theglobeandmail.com/business/article-british-columbia-mining-cleanup-costs-investigation.

33 Eliane Brum, *Banzeiro: The Amazon as the Centre of the World,"* trans. Diane Whitty (Minneapolis: Graywolf Press, 2023), 22.

34 Angele Alook, Emily Eaton, David Gray-Donald, Joël Laforest, Crystal Lameman, and Bronwen Tucker (eds.), *The End of This World: Climate Justice in So-Called Canada* (Toronto: Between the Lines, 2023).

35 Alook et al., *The End of This World,* 10.

36 VersoBooks, "What is Ecofeminism, and Why Is It Necessary in the Fight for Climate Justice?" *Climate Crisis: Time for a New Society* (podcast), https://soundcloud.com/versobooks/what-is-ecofeminism-and-why-is-it-necessary-in-the-fight-for-climate-justice.

37 Dian Million, *Therapeutic Nations: Healing in an Age of Indigenous Human Rights* (Tuscon: Arizona University Press, 2014); see also Dian Million, "Felt Theory: An Indigenous Feminist Approach to Affect and History," *Wicazo Sa Review* 24, 2 (2009): 53–76.

38 For more about the Feminist Environmental Research Network see www.ferncollaborative.com

39 Alys Einion and Jen Rinaldi (eds.), *Bearing the Weight of the World: Exploring Maternal Embodiment* (Bradford: Demeter Press, 2018).

40 bell hooks, *all about love* (New York: Harper-Collins, 2018).

41 bell hooks, *Feminism Is for Everybody* (London: Pluto Press, 2000).

CODE RED: FEMINIST MOTHERHOOD IN A WORLD ON FIRE

1. Elizabeth Weil, "This Isn't the California I Married," *New York Times*, January 3, 2022, https://www.nytimes.com/2022/01/03/magazine/california-widfires.html.
2. BC Women's Hospital and Health Centre, "How to Breathe Easier during Wildfire Season when Pregnant," June 13, 2023.
3. Dawn Hoogeveen, Kerri Klein, Jordan Brubacher, and Maya K. Gislason, *Climate Change, Intersectionality and GBA+ in British Columbia*, Summary Report prepared for the Climate Action Secretariat, March 2021, https://www2.gov.bc.ca/assets/gov/environment/climate-change/adaptation/resources/climate_change__gba_in_bc_summary_report.pdf; Olena Hankivsky and Julia S. Jordan-Zachery, *The Palgrave Handbook of Intersectionality in Public Policy* (London: Palgrave Macmillan, 2019).
4. Anelyse M. Weiler, "Farmworkers, Climate Change, and 'Converging Crises,'" *Gastronomica* 22, 1 (Spring 2022): 44–49.
5. Weiler, "Farmworkers, Climate Change, and 'Converging Crises.'"
6. Weiler, "Farmworkers, Climate Change, and 'Converging Crises.'"
7. The Editorial Board, "Scenes from a World on Fire," *New York Times*, December 31, 2021, https://www.nytimes.com/2021/12/31/opinion/climate-change-glasgow-united-states.html.
8. Sarah Marie Wiebe, "Sensing Policy: Engaging Affected Communities at the Intersections of Environmental Justice and Decolonial Futures," *Politics, Groups and Identities* 8, 1 (2020): 181–193.
9. Kathryn Blaze Baum and Matthew McClearn, "Extreme, Deadly Heat in Canada Is Going to Come Back, Worse Than Ever. Will We Be Ready?" *Globe and Mail*, September 21, 2021, https://www.theglobeandmail.com/canada/article-extreme-deadly-heat-in-canada-is-going-to-come-back-and-worse-will-we/.
10. Henry Fountain, "The Vast Wildfire Lab Is Helping Foresters to Prepare for a Hotter Planet," *New York Times*, January 5, 2022, https://www.nytimes.com/2022/01/05/climate/fire-forest-management-bootleg-oregon.html.
11. Chad Pawson, "Climate Change Disasters in B.C. Likely to Increase

if Industrial Logging Continues Unchecked: Report," *CBC News*, February 1, 2021, https://www.cbc.ca/news/canada/british-columbia/logging-climate-change-bc-report-1.5895220. See also Peter Wood, "Intact Forests, Safe Communities: Reducing Community Climate Risks Through Forest Protection and a Paradigm Shift in Forest Management," *Sierra Club BC,* February 2021, https://sierraclub.bc.ca/intact-forests-safe-communities-sierra-club-bc-report/.

12 For more about our research see the story map online: "A Hot Topic: Understanding Heat Vulnerability in the Capital Regional District," https://storymaps.com/stories/989b03cc3bc042fa8d7fc6e80c712464 and also the 2023 research report available online: Sarah Marie Wiebe and Kirsten Mah, *A Hot Topic: Feeling the Impacts of Extreme Heat,* https://www.sarahmariewiebe.com/ahottopic.

13 Hawai'i State Commission on the Status of Women, Department of Human Services, State of Hawai'i, *Building Bridges, Not Walking on Backs: A Feminist Economic Recovery Plan for COVID-19.* See also Nancy Fraser, "Contradictions of Capital and Care," *New Left Review,* July/August 2016, https://newleftreview.org/issues/ii100/articles/nancy-fraser-contradictions-of-capital-and-care.

14 Hawai'i State Commission on the Status of Women, *Building Bridges*, 3.

15 Marina Romanello, Alice McGushin, Claudia Di Napoli, Paul Drummond, Nick Hughes, Louis Jamart, et al., "The 2021 Report of the *Lancet* Countdown on Health and Climate Change: Code Red for a Healthy Future," *Lancet* 398, 10311 (2021): 619–1662, https://www.thelancet.com/journals/lancet/article/PIIS0140-6736(21)01787-6/fulltext.

16 Rhianna Schmunk, "In a Single Week: A Timeline of How Once-in-a-Century Flooding Unfolded across B.C." *CBC News,* November 20, 2021, https://newsinteractives.cbc.ca/longform/bc-flooding-2021-timeline-how-once-in-a-century-flooding-unfolded/.

17 Intergovernmental Panel on Climate Change (IPCC), *Climate Change 2021: The Physical Science Basis,* 6th Assessment Report, August 6, 2021, https://www.ipcc.ch/report/ar6/wg1/.

18 United Nations News. "IPCC Report: 'Code Red' for Human Driven Global Heating, Warns UN Chief," August 2021, https://news.un.org/en/story/2021/08/1097362.

19 United Nations News, "UN Secretary-General: COP26 Must Keep 1.5 Degrees Celsius Goal Alive," November 1, 2021, https://unfccc.int/news/un-secretary-general-COP26-must-keep-15-degrees-celsius-goal-alive?utm_source=BenchmarkEmail&utm_campaign=CED-Nov-2021&utm_medium=email.

20 Brad Plumer and Elena Shao, "Heat Records Are Broken around the Globe as the Earth Warms, Fast," *New York Times,* July 6, 2023, https://www.nytimes.com/2023/07/06/climate/climate-change-record-heat.html.

21 Winston Choi-Schagrin, "New Research Shows How Health Risks to Children Mount as Temperatures Rise," *New York Times,* January 19, 2022, https://www.nytimes.com/2022/01/19/climate/children-climate-change.html.

22 Brian P. Dunleavy, "Extreme Heat Raises Risk for Mental Health Crises in U.S., Study Finds," *UPI Health News,* February 23, 2022, https://www.upi.com/Health_News/2022/02/23/extreme-heat-mental-health-risk-study/9321645625261/.

23 Maria Cramer, "Baby Born 19 Weeks Early Defies Long Odds and Astonishes Doctors," *New York Times* November 13, 2021, https://www.nytimes.com/2021/11/13/us/premature-baby-guinness-record.html.

24 Paola Scommenga, "High Premature Birth Rates Among U.S. Black Women May Reflect the Stress of Racism and Health and Economic Factors," *Population Reference Bureau,* January 21, 2021, https://www.prb.org/resources/high-premature-birth-rates-among-u-s-black-women-may-reflect-the-stress-of-racism-and-health-and-economic-factors/.

25 Christopher Flavelle, "Climate Change Tied to Pregnancy Risks, Affecting Black Mothers Most," *New York Times,* June 18, 2020, https://www.nytimes.com/2020/06/18/climate/climate-change-pregnancy-study.html.

26 For more on the politics of care, see the 2021 Critical Exchange in *Contemporary Political Theory:* Deva Woodly, Rachel H. Brown, Mara Marin, Shatema Threadcraft, Christopher Paul Harris, Jasmine Syedullah, and Miriam Ticktin, "The Politics of Care," *Contemporary Political Theory* 20, 4 (2021): 890–925. See also a forthcoming Critical Exchange in *Contemporary Political Theory* on Feminist Approaches to Environmental Politics, by Isabel Altamirano-Jiménez, Cara Daggett, Jennifer L. Lawrence, Sherilyn MacGregor, Emily Ray, Sarah Marie Wiebe, Hannah Battersby, Magdalena Rodekirchen, and Heather Urquhart.

27 Nina Lakhani, "'A Continuation of Colonialism': Indigenous Activists Say Their Voices Are Missing from COP26," *Guardian,* November 3, 2021, https://www.theguardian.com/environment/2021/nov/02/COP26-indigenous-activists-climate-crisis.

CODE ORANGE: CULTIVATING COMMUNITY THROUGH DISASTERS

1 The impact of Canada's Chemical Valley on the community health of the Aamjiwnaang Nation is discussed at length in my book. See Sarah Marie Wiebe, *Everyday Exposure: Indigenous Mobilization and Environmental Justice in Canada's Chemical Valley* (Vancouver: UBC Press, 2016). For an elaboration of concerns related to endocrine disruption, pollution, and culture, see information available through the Endocrine Disruptors Action Group website: https://endocrinedisruptorsaction.org/. Ongoing community-engaged research about these intersecting impacts and relationships is being led by Michelle Murphy and the Environmental Data Justice Lab. See https://technoscienceunit.org/people/lab/.

2 Sarnia Historical Society, "Polymer and the War Effort: 1942," https://www.sarniahistoricalsociety.com/story/polymer-and-the-war-effort-1942/.

3 For a detailed discussion, see my book, *Everyday Exposure*.

4 See the website of Clean Air Sarnia and Area (CASA): https://www.cleanairsarniaandarea.com/reporting/station-dedications.aspx.

5 For a discussion of the activism behind victims of Chemical Valley, see Maia Foulis, "In Chemical Valley, Widows Have Waited Decades

for Compensation," *Canadian Occupational Safety,* January 24, 2022, https://www.thesafetymag.com/ca/topics/occupational-hygiene/in-chemical-valley-widows-have-waited-decades-for-compensation/322895.

6 Ada explains these multijurisdictional configurations in her poem "I Didn't Know," cited in *Toxic News,* February 21, 2019, https://toxicnews.org/2019/02/21/bodies-exposed-reframing-the-geopolitics-of-dilution-in-canadas-chemical-valley/.

7 Constanze A. Mackenzie, Ada Lockridge, and Margaret Keith, "Declining Sex Ratio in a First Nation Community," *Environmental Health Perspectives* 113, 10 (2005): 1295–1298.

8 Wiebe, *Everyday Exposure.*

9 Rob Nixon, *Slow Violence and the Environmentalism of the Poor* (Cambridge, MA: Harvard University Press, 2011).

10 Winston Choi-Schagrin, "Sometimes, Life Stinks. So He Invented the Nasal Ranger," *New York Times,* January 13, 2022, https://www.nytimes.com/2022/01/13/climate/nasal-ranger-chuck-mcginley.html.

11 Jennifer Lawrence and Sarah Marie Wiebe, *Biopolitical Disaster* (London and New York: Routledge, 2017).

12 Leslie Brown and Susan Strega (eds.), *Research as Resistance* (Toronto: Canadian Scholars' Press, 2015); Margaret Kovach, "Emerging from the Margins: Indigenous Methodologies," in Brown and Strega, *Research as Resistance*, 43–64; Margaret Kovach, *Indigenous Methodologies: Characteristics, Conversations and Contexts* (Toronto: University of Toronto Press, 2009); Linda Tuhiwai Smith, *Decolonizing Methodologies: Research and Indigenous Peoples* (London: Zed Books, 2012).

13 Jennifer Lawrence, Emily Ray, and Sarah Marie Wiebe, "Critical Ecofeminism: A Feminist Environmental Research Network (FERN) for Collaborative and Relational Praxis," in *The Palgrave Handbook of Environmental Politics and Theory,* eds. Joel Kassiola and Timothy W. Luke (Palgrave Macmillan, 2023) 195–222; see also *Feminist Approaches to Environmental Politics*, forthcoming Critical Exchange in *Contemporary Political Theory.*

14 For some inspiring and imaginative thinkers at the nexus of feminist biopolitics, geopolitics, and affect theory, see Stefanie Fishel, *The Microbial State: Global Thriving and the Body Politic* (Minneapolis: University of Minnesota Press, 2017); Dian Million, *Therapeutic Nations: Healing in an Age of Indigenous Human Rights* (Tucson: University of Arizona Press, 2013); Ben Anderson, "Affective Atmospheres," *Emotion, Space and Society* 2 (2009): 77–81; Ben Anderson, *Encountering Affect* (London: Routledge, 2014).

CODE PINK: A SPLASH AND DASH CAESAREAN

1 Dawn Marie Smith, "hiiɬ kʷiiʔiɬ siλ (Bringing Something Good from Way Back): A Journey to Humanize Post-Secondary Education" (doctoral dissertation, University of British Columbia, 2018), https://open.library.ubc.ca/soa/cIRcle/collections/ubctheses/24/items/1.0371220.

2 I cite the Nuu-chah-nulth teaching of yaʔakmis because its emphasis on love and pain oriented me during a particularly difficult birthing and perinatal experience. Ethical citation of this family cultural teaching in settler-authored scholarship requires that I do not decontextualize it or imply an authoritative interpretation of it.

3 Dawn Marie Smith, "hiiɬ kʷiiʔiɬ siλ," 80. See also Richard Atleo, *Tsawalk: A Nuu-Chah-Nulth Worldview* (Vancouver: UBC Press, 2004).

4 Editorial Board, Opinion, "Scenes from a World on Fire," *New York Times*, December 31, 2021, https://www.nytimes.com/2021/12/31/opinion/climate-change-glasgow-united-states.html; William J Ripple, Christopher Wolf, Jillian W Gregg, Kelly Levin, Johan Rockström, Thomas M. Newsome, Matthew G. Betts, Saleemul Huq, Beverly E. Law, Luke Kemp, Peter Kalmus, and Timothy M. Lenton "World Scientists' Warning of a Climate Emergency," *BioScience* 72, 12 (2022): 1149–1155.

5 See postcard visualizations of a world on fire entitled *Cities Swallowed by Dust* in *New York Times*, interactive essay, December 13, 2021, https://www.nytimes.com/interactive/2021/12/13/opinion/climate-change-effects-countries.html.

6 Interview with Katsi Cook in *Braided Way: Faces and Voices of Spiritual Practice,* November 3, 2018, https://braidedway.org/women-are-the-first-environment-interview-with-mohawk-elder-katsi-cook/.

7 Kim Smith, "Why Is the C-Section Rate Still Climbing in Canada?" *Global News,* May 28, 2019, https://globalnews.ca/news/5325680/family-matters-c-section-rate-climbing-canada-birth-health/.

8 Tabitha Soren, "The Blurred Existence of Motherhood," *The Atlantic,* May 9, 2021, https://www.theatlantic.com/magazine/archive/2021/06/tabitha-soren-motherload/618718/.

9 Sara Ahmed, *The Cultural Politics of Emotion* (New York: Routledge, 2015), 38.

10 Alys Einion and Jen Rinaldi, *Bearing the Weight of the World: Exploring Maternal Embodiment* (Bradford: Demeter Press, 2018), 11.

CODE BLUE: LIVING THROUGH MULTIPLE CRISES, CLIMATE ANXIETY, AND MENTAL HEALTH

1 Environmental Protection Agency, *Climate Change and the Health of Pregnant, Breastfeeding, and Postpartum Women,* updated December 27, 2023, https://www.epa.gov/climateimpacts/climate-change-and-health-pregnant-breastfeeding-and-postpartum-women.

2 For more about our "Hot Topic" research project, see our story map available online: "A Hot Topic: Understanding Heat Vulnerability in the Capital Regional District," https://storymaps.com/stories/989b03cc3bc042fa8d7fc6e80c712464.

3 Alex Migdal, "More Than a Billion Seashore Animals May Have Cooked to Death in B.C. Heat Wave, Says UBC Researcher," *CBC News,* July 5, 2021, https://www.cbc.ca/news/canada/british-columbia/intertidal-animals-ubc-research-1.6090774.

4 Leyland Cecco, "'Heat Dome' Probably Killed 1bn Marine Animals on Canada Coast, Experts Say," *Guardian,* July 8, 2021, https://www.theguardian.com/environment/2021/jul/08/heat-dome-canada-pacific-northwest-animal-deaths.

5 *CBC News,* "B.C. Park Name Change Recognizes Ancestral Home of the Tsleil-Waututh Nation," October 9, 2021, https://www.cbc.ca/

news/canada/british-columbia/t%C9%99mt%C9%99m%C3%ADx%CA%B7t%C9%99n-belcarra-regional-park-1.6205806.

6 Rachel yacaaʔaɬ George and Sarah Marie Wiebe, "Fluid Decolonial Futures: Water as Life, Ocean Citizenship and Seascape Relationality," *New Political Science* 42, 4 (2020): 498–520. https://doi.org/10.1080/07393148.2020.1842706

7 Kevin Hamilton, "Tonga Eruption Was So Intense, It Caused the Atmosphere to Ring Like a Bell," *Civil Beat,* January 27, 2022, https://www.civilbeat.org/?p=1477302&utm_source=Civil+Beat+Master+List&utm_campaign=8145d5cb90-EMAIL_CAMPAIGN_2022_01_27_05_41&utm_medium=email&utm_term=0_51c2dd3cf3-8145d5cb90-401791737&mc_cid=8145d5cb90&mc_eid=8158a6090a.

8 Henry Fountain, "Here's What Scientists Know About the Tonga Volcano Eruption," *New York Times*, January 19, 2022, https://www.nytimes.com/2022/01/19/climate/scientists-tonga-volcano-eruption-effects.html?campaign_id=54&emc=edit_clim_20220119&instance_id=50761&nl=climate-fwd%3A&regi_id=81926201&segment_id=80173&te=1&user_id=66093555d645d88994ae4209f7d9a81c.

9 National Collaborating Centre for Environmental Health, *Overview of Canadian Communities Exposed to Sea Level Rise,* December 7, 2022, https://ncceh.ca/resources/evidence-reviews/overview-canadian-communities-exposed-sea-level-rise#.

10 Henry Fountain, "Coastal Sea Levels in U.S. to Rise a Foot by 2050, Study Confirms," *New York Times,* February 15, 2022, https://www.nytimes.com/2022/02/15/climate/us-rising-sea-levels.html.

11 Next Generation Radio and Claire Caulfield, "From Fishermen to Rugby Players, Climate Change Has Become a Fact of Life in Fiji," *Honolulu Civil Beat*, Special Project, February 17, 2022, https://www.civilbeat.org/?p=1480493&utm_source=Civil+Beat+Master+List&utm_campaign=062178dddf-EMAIL_CAMPAIGN_2022_02_17_05_01&utm_medium=email&utm_term=0_51c2dd3cf3-062178dddf-401791737&mc_cid=062178dddf&mc_eid=8158a6090a.

12 Matthew McClearn, "Canada's Disappearing Coastline: How Climate Change Puts Our Beaches in Jeopardy," *Globe and Mail,* August 20, 2021, https://www.theglobeandmail.com/canada/article-canadas-disappearing-coastline-how-climate-change-puts-our-sandy/.

13 Jasper Knight, "World's Beaches Are Changing Because of Climate Change: Green Thinking Is Needed to Save Them," *The Conversation,* August 24, 2023, https://theconversation.com/worlds-beaches-are-changing-because-of-climate-change-green-thinking-is-needed-to-save-them-211953.

14 Aidan Gardiner, "Disasters Forced 2.5 Million Americans from Their homes Last Year," *New York Times.* February 22, 2024, https://www.nytimes.com/2024/02/22/climate/climate-disasters-survivors-displacement.html.

15 Next Generation Radio and Caulfield, "From Fishermen to Rugby Players."

16 Rhianna Schmunk, "In a Single Week: A Timeline of How a Once-in-a-Century Flooding Unfolded across B.C.," CBC *News,* November 20, 2021, https://newsinteractives.cbc.ca/longform/bc-flooding-2021-timeline-how-once-in-a-century-flooding-unfolded.

17 What on Earth, CBC Explains, "What are Atmospheric Rivers, and How are They Affecting the B.C. Floods?" November 18, 2021, https://www.cbc.ca/radio/whatonearth/what-are-atmospheric-rivers-and-how-are-they-affecting-the-b-c-floods-1.6253763.

18 Glenn McGillivray and Korey Pasch, "Disasters are a Permanent Part of the Canadian Fabric. We Can't Keep Winging It," *Globe and Mail,* December 6, 2021, https://www.theglobeandmail.com/opinion/article-disasters-are-a-permanent-part-of-the-canadian-fabric-we-cant-keep/.

19 Netanya Castillo, "It's a Perfect Storm of Stress. How to Cope," *The Tyee,* November 30, 2021, https://thetyee.ca/News/2021/11/30/Perfect-Storm-Stress-How-Cope/.

20 IPCC, *Synthesis Report of the IPCC Sixth Assessment Report (AR6),* 2023, 6, https://report.ipcc.ch/ar6syr/pdf/IPCC_AR6_SYR_SPM.pdf.

21 Kassandra Kometani, "Environmental Justice for Native Hawaiians:

Preventing Landfill Expansion on the Waiʻanae Coast," *Environmental Law Education Centre*, March 3, 2022, https://elecenter.com/1143/environmental-justice-for-native-hawaiians-preventing-landfill-expansion-on-the-waianae-coast/.

22 Kuʻu Kuanoe, "From 'Sacred Place' to 'Dumping Ground,' West Oahu Confronts a Legacy of Landfills," *Civil Beat,* December 5, 2021, https://www.civilbeat.org/?p=1469558&utm_source=Civil+Beat+Master+List&utm_campaign=ec24089e73-EMAIL_CAMPAIGN_2021_12_05_07_12&utm_medium=email&utm_term=0_51c2dd3cf3-ec24089e73-401791737&mc_cid=ec24089e73&mc_eid=8158a6090a/.

23 Farhana Sultana, "The Unbearable Heaviness of Climate Coloniality," *Political Geography* 99 (November 2022), https://www.sciencedirect.com/science/article/pii/S096262982200052X.

24 World Health Organization and Calouste Gulbenkian Foundation, *Social Determinants of Mental Health* (Geneva: World Health Organization, 2014), https://apps.who.int/iris/bitstream/handle/10665/112828/9789241506809_eng.pdf.

25 Makshya Tolbert, "Becoming Water: Black Memory in Slavery's Afterlives," *Emergence Magazine,* February 17, 2022, https://emergencemagazine.org/essay/becoming-water/.

26 Eduardo Medina and Alyssa Lukpat, "Flash Floods Hit Parts of Hawaii as Storm Lashes Region," *New York Times,* December 6, 2021, https://www.nytimes.com/2021/12/06/us/hawaii-flooding.html.

27 Cammy Clark, "Lahaina Fire Death Toll Rises to 101 after Police Identify Remains of Missing Person," *Civil Beat,* February 13, 2024, https://www.civilbeat.org/2024/02/lahaina-fire-death-toll-rises-to-101-after-police-identify-remains-of-missing-person/; see also Reuters, "Deadliest US Wildfires," https://www.reuters.com/graphics/HAWAII-WILDFIRES/DEATHS/lbpgoonxapq/index.html?te=1&nl=david-wallace-wells&emc=edit_dww_20240220.

28 Julian Augon, "To Hell with Drowning," *The Atlantic,* November 1, 2021, https://www.theatlantic.com/culture/archive/2021/11/oceania-pacific-climate-change-stories/620570/.

29 Epeli Hauʻofa, "Our Sea of Islands," *The Contemporary Pacific* 6, 1 (1994): 148–161.
30 Kathy Jetñil-Kijiner, "Anointed," April 16, 2018, YouTube video, 6:08, https://www.youtube.com/watch?v=HuDA7izeYrk/.
31 Reuben George with Mike Simpson, *It Stops Here: Standing Up for Our Lands, Our Waters and Our People* (Toronto: Penguin Random House Canada, 2023).
32 Erin Anderssen, "Our Brains Don't Like Uncertainty, But Research Offers Clues for How to Cope," *Globe and Mail*, December 5, 2021, https://www.theglobeandmail.com/canada/article-our-brains-dont-like-uncertainty-but-research-offers-clues-for-how-to/.

CODE GREEN: CIRCULAR ECONOMIES OF CARE

1 Don Descoteau, "Victoria Lacks More Than 4,200 Child Care Spaces within City: Report," *Saanich News*, October 6, 2020, https://www.saanichnews.com/news/victoria-lacks-more-than-4200-child-care-spaces-within-city-report-291279.
2 Descoteau, "Victoria Lacks More Than 4,200 Child Care Spaces."
3 Angele Alook, Emily Eaton, David Gray-Donald, Joël Laforest, Crystal Lameman, and Bronwen Tucker, *The End of This World: Climate Justice in So-Called Canada* (Toronto: Between the Lines, 2023), 111.
4 Alook et al., *The End of This World*, 114.
5 Deva Woodly, Rachel H. Brown, Mara Marin, Shatema Threadcraft, Christopher Paul Harris, Jasmine Syedullah, and Miriam Ticktin, "The Politics of Care," *Contemporary Political Theory* 20, 4 (2021): 890–925. See also Rebecca Jane Hall, *Refracted Economies: Diamond Mining and Social Reproduction in the North* (Toronto: University of Toronto Press, 2022).
6 Nancy Fraser, "Contradictions of Capital and Care," *New Left Review*. July/August 2016, https://newleftreview.org/issues/ii100/articles/nancy-fraser-contradictions-of-capital-and-care.
7 Alook et al., *The End of This World*, 123–134.
8 Josie Kao, "Canadian Fashion Retailers Leading the Way in

Sustainability," *Globe and Mail,* November 2, 2021, https://www.theglobeandmail.com/business/article-canadian-fashion-retailers-seek-sustainability-practices-that-are-the/.

9   Kamana Beamer, "Following the Principles of Aloha Aina — Respect, Reverence, Justice — Will Be Instrumental to Our Well-Being in the Years Ahead," *Civil Beat,* August 1, 2021, https://www.civilbeat.org/2021/07/give-take-regenerate-a-circular-economy-is-vital-to-hawaiis-future/.

10 Nancy Fraser, "Contradictions of Capital and Care," *New Left Review* 100 (2016): 99–117, https://newleftreview.org/issues/ii100/articles/nancy-fraser-contradictions-of-capital-and-care.

11 For further reading on this topic, see Dr. Ethel Tungohan's work, including *Care Activism: Migrant Workers, Movement Building and Communities of Care* (Chicago: University of Illinois Press, 2023).

12 Fraser, "Contradictions of Capital and Care."

13 See my article co-authored with Lynne Ng, "Reproductive Futurism through Intergenerational Care," under review for *The Politics of Reproduction* edited collection.

14 Read about planetary health in the Association of Faculties of Medicine of Canada's Academic Institutions' Declaration on Planetary Health online: https://www.afmc.ca/initiatives/planetaryhealthdeclaration/.

15 Rachel yacaaʔał George and Sarah Marie Wiebe, "Fluid Decolonial Futures: Water as a Life, Ocean Citizenship and Seascape Relationality," *New Political Science* 42, 4 (2020): 498–520.

16 Kamanamaikalani Beamer, "Only Twenty Ahupuaʻa Away," in *Detours: A Decolonial Guide to Hawaiʻi,* eds. Hōkūlani K. Aikau and Vernadette Vicuña Gonzalez (Durham: Duke University Press, 2019), 21.

17 Melaine Haiken, "Hawaii's Ancient Land Management System," BBC *Travel,* August 18, 2022, https://www.bbc.com/travel/article/20220818-ahupuaa-hawaiis-ancient-land-management-system.

18 See the online Hawaiian dictionary at https://wehewehe.org/.

19 Marc Lee and Belinda Li, "Getting to Zero Waste by 2040," *The Tyee,* January 21, 2022, https://thetyee.ca/Opinion/2022/01/21/Getting-Zero-Waste-2040/.

20 Beamer, "Only Twenty Ahupuaʻa Away," 24.

21 ʻĀina Aloha Economic Futures, https://www.ainaalohafutures.com.

22 As explained, for example, by Carol Anne Hilton in her work on Indigenomics. See Carol Anne Hilton, *Indigenomics: Taking a Seat at the Economic Table* (Gabriola, BC: New Society Publishers, 2021).

23 ʻAina Aloha Economic Futures, *Growing a Stronger Hawaiʻi: Policy Playbook*, https://docs.google.com/document/d/1GJdVPM84fA0x9UBGpii0UjBgvEiV4INr2ES5FXmRpjk/edit?usp=embed_facebook.

24 Alook et al., *The End of This World*, 128.

25 Lisa Abend, "Inside Finland's Plan to End All Waste by 2050," *TIME Magazine*, January 20, 2022, https://time.com/6132391/finland-end-waste/.

26 See Recognizing the Duty of the Federal Government to Create a Green New Deal, H.Res.109., 116th Congress (2019–2020), https://www.congress.gov/bill/116th-congress/house-resolution/109/text.

27 *The Red Nation: Decolonization or Extinction*, "The Red Deal: Indigenous Action to Save Our Earth" (podcast), https://therednation.org/the-red-deal/.

28 WECAN (Women's Earth & Climate Action Network) "WECAN at UNFCCC COP28," https://www.wecaninternational.org/cop28.

29 FemGND, "Reflection Points on COP26 – Feminist Agenda for a Green New Deal" (blog), December 20, 2021, http://feministgreennewdeal.com/2021/12/20/reflection-points-on-cop26/.

30 FemGND, "Reflection Points on COP26."

31 WECAN "WECAN at UNFCCC COP28."

32 Hall, *Refracted Economies*, 224.

33 Ingrid Waldron, *There's Something in the Water: Environmental Racism in Indigenous and Black Communities* (Halifax: Fernwood Publishing, 2018).

34 EJAtlas, "Canada," *Global Atlas of Environmental Justice*, https://ejatlas.org/country/canada.

35 Hawaiʻi State Commission on the Status of Women, Department of Human Services, State of Hawaiʻi, *Building Bridges, Not Walking on*

Backs: A Feminist Economic Recovery Plan for COVID-19,
https://humanservices.hawaii.gov/wp-content/uploads/2020/04/
4.13.20-Final-Cover-D2-Feminist-Economic-Recovery-D1.pdf.

36 Teena Gabrielson and Katelyn Parardy, "Corporeal Citizenship: Rethinking Green Citizenship through the Body," *Environmental Politics* 19, 3 (2010): 380.

37 For example, see Jennifer Lawrence, Emily Ray, and Sarah Marie Wiebe, "Critical Ecofeminism: A Feminist Environmental Research Network (FERN) for Collaborative and Relational Praxis," in *The Palgrave Handbook of Environmental Politics and Theory,* eds. Joel Kassiola and Timothy W. Luke (London: Palgrave Macmillan, 2023), 195–222; see also *Feminist Approaches to Environmental Politics*, forthcoming Critical Exchange in *Contemporary Political Theory.*

38 Gabrielson and Parardy, "Corporeal Citizenship," 382.

39 Thank-you to Tanya Andrusieczko for this generative exchange, which deepened my thinking about the treaty-like sensations of nursing. See also Interview with Katsi Cook in *Braided Way: Faces and Voices of Spiritual Practice,* November 3, 2018, https://braidedway.org/women-are-the-first-environment-interview-with-mohawk-elder-katsi-cook/.

40 Gabrielson and Parardy, "Corporeal Citizenship," 382.

41 Gabrielson and Parardy, "Corporeal Citizenship," 387.

CODE BLACK: SYSTEMIC THREATS, REVEALING VIOLENCES SLOW AND SPECTACULAR

1 Hōkulani K. Aikau and Vernadette Vicuña Gonzalez (eds.), *Detours: A Decolonial Guide to Hawai'i* (Durham: Duke University Press, 2019), 1–2.

2 *On the Morning You Wake (To the End of the World)* (webpage), https://www.onthemorningyouwake.com/

3 *On the Morning You Wake,* "About," https://www.onthemorningyouwake.com/#about

4 Aikau and Gonzalez (eds.), *Detours,* 9.

5   Aikau and Gonzalez (eds.), *Detours*, 5. See also Noelani Goodyear-Ka'ōpua, Ikaika Hussey, and Erin Kahunawaika'ala Wright (eds.), *A Nation Rising: Hawaiian Movements for Life, Land and Sovereignty* (Durham: Duke University Press, 2014).

6   Anita Hofschneider, "'The Stakes Are High': Why the Navy Doesn't Want to Defuel the Red Hill Tanks," *Civil Beat,* December 22, 2021, https://www.civilbeat.org/?p=1473051&utm_source=Civil+Beat+Master+List&utm_campaign=c8e3499fd9-EMAIL_CAMPAIGN_2021_12_22_07_51&utm_medium=email&utm_term=0_51c2dd3cf3-c8e3499fd9-401791737&mc_cid=c8e3499fd9&mc_eid=8158a6090a; Sarah Marie Wiebe, "Sensing Empire at Sea: SONAR, Kanaloa and Indigenous Marine Sovereignty," *Sensaste: A Journal for Experiments in Critical Media Practice* (2021), https://sensatejournal.com/sensing-empire-at-sea-sonar-kanaloa-and-indigenous-marine-sovereignty/.

7   Nick Grube, "Hawaii Activists Travel to DC to Demand Action on Red Hill," *Civil Beat,* September 20, 2022, https://www.civilbeat.org/beat/hawaii-activists-travel-to-dc-to-demand-action-on-red-hill/.

8   Sterling Higa, "Sterling Higa: What Red Hill Says about the Decline of American Empire," *Civil Beat,* December 22, 2021, https://www.civilbeat.org/?p=1472841&utm_source=Civil+Beat+Master+List&utm_campaign=c8e3499fd9-EMAIL_CAMPAIGN_2021_12_22_07_51&utm_medium=email&utm_term=0_51c2dd3cf3-c8e3499fd9-401791737&mc_cid=c8e3499fd9&mc_eid=8158a6090a.

9   Democracy Now, "'Shut Down Those Tanks': Anger Grows in Hawaii after U.S. Navy Fuel Site Contaminates Water," *Democracy Now!: Independent Global News,* December 21, 2021, https://www.democracynow.org/2021/12/21/petroleum_leak_contaminates_water_hawaii.

10 Lauren Teruya, "The Navy Illegally Flushed Tainted Water Into Storm Drains as Red Hill Crisis Took Hold," *Civil Beat,* December 22, 2021, https://www.civilbeat.org/?p=1471517&utm_source=Civil+Beat+Master+List&utm_campaign=c8e3499fd9-EMAIL_CAMPAIGN_2021_12_22_07_51&utm_medium=email&utm_

term=0_51c2dd3cf3-c8e3499fd9-401791737&mc_cid=c8e3499fd9&mc_eid=8158a6090a.

11 Sierra Club of Hawai'i, "Shut Down Red Hill," https://sierraclubhawaii.org/redhill-1.

12 Megan K. Stack, "U.S. Soldiers Came Home Sick. The Government Denied Responsibility," *New York Times*, January 11, 2022, https://www.nytimes.com/2022/01/11/magazine/military-burn-pits.html.

13 Ingrid Waldron. *There's Something in the Water: Environmental Racism in Indigenous and Black Communities* (Halifax: Fernwood Publishing, 2018); The ENRICH Project: The Environmental Noxiousness, Racial Inequities & Community Health Project, online at https://www.enrichproject.org.

14 Mapping Police Violence initiative, updated March 20, 2024, https://mappingpoliceviolence.org.

15 Richard Meadow, "Police Brutality Statistics: What the Data Says about Police Violence in America," Police Brutality Centre, June 27, 2022, https://policebrutalitycenter.org/police-brutality/statistics/.

16 Debra Thompson, *The Long Road Home: On Blackness and Belonging* (New York: Simon & Schuster, 2022).

17 Waldron. *There's Something in the Water*, 9.

18 Susan Goldberg, "Inside the Capitol: 'It was the Ugliest Moment I Have Ever Seen in America,'" *National Geographic*, January 7, 2021, https://www.nationalgeographic.com/photography/article/louie-palu-interview-inside-the-capitol.

19 Ian Bailey, "Emergencies Act Will Stay in Place Because of Continued Threat Linked to Protests, Trudeau Says," *Globe and Mail*, February 21, 2022, https://www.theglobeandmail.com/politics/article-politics-briefing-emergencies-act-will-stay-in-place-because-of/.

20 Joyce Green and Gina Starblanket, "Law, Order, Emergencies and Enforcement: Who Matters?" *The Tyee*, February 16, 2022, https://thetyee.ca/Analysis/2022/02/16/Law-Order-Emergencies-Enforcement/.

21 Sarah Marie Wiebe, *Life against States of Emergency: Revitalizing Treaty Relations from Attawapiskat* (Vancouver: UBC Press, 2023).

22 Judith Butler, *The Force of Nonviolence: An Ethico-Political Bind* (London: Verso, 2020), 73

23 Butler, *The Force of Nonviolence*, 147.

24 Somini Sengupta, "Chile Writes a New Constitution, Confronting Climate Change Head On," *New York Times,* December 28, 2021, https://www.nytimes.com/2021/12/28/climate/chile-constitution-climate-change.html.

25 A compelling term coined by authors Alana Cattapan, April Mandrona, Tammy Findlay, and Alexandra Dobrowolsky, "Power, Privilege, and Policy Making: Reflections on Changing Public Engagement from the Ground Up," in *Creating Spaces of Engagement: Policy Justice and the Practical Craft of Deliberative Democracy*, eds. Leah Levac and Sarah Marie Wiebe (Toronto: University of Toronto Press, 2020), 226–252.

26 bell hooks, *all about love* (New York: Harper-Collins, 2018), 129.

27 hooks, *all about love*, 19.

28 hooks, *all about love*, 30.

29 hooks, *all about love*, 87.

30 hooks, *all about love,* 87-88.

31 hooks, *all about love*, 161

32 hooks, *all about love*, 162

33 hooks, *all about love,* 133-134.

34 hooks, *all about love*, 136.

35 hooks, *all about love,* 144.

CODE GREY: A CAUTIONARY TALE OF RENEWABLE EXTRACTION

1 See also a detailed discussion in the film by Vicki Lean (director), *The Climate Baby Dilemma,* CBC, 2022.

2 United Nations, "Renewable Energy: Powering a Safer Future," *Climate Action,* https://www.un.org/en/climatechange/raising-ambition/renewable-energy.

3 Christina Nunez, "Renewable Energy, Explained," *National Geographic,* January 30, 2019, https://www.nationalgeographic.com/environment/article/renewable-energy.

4 Konrad Yakabuski, "Transition to EVs Could Be Only Hope for Canada's Shrinking Auto Industry," *Globe and Mail,* February 2, 2022, https://www.theglobeandmail.com/business/commentary/article-transition-to-evs-could-be-only-hope-for-canadas-shrinking-auto/.

5 Andrew Nikiforuk, "Are Electric Cars the Solution?" *The Tyee,* January 25, 2022, https://thetyee.ca/Analysis/2022/01/25/Are-Electric-Cars-Solution/.

6 Dionne Searcey, Michael Forsythe, and Eric Lipson, "A Power Struggle over Cobalt Rattles the Clean Energy Revolution," *New York Times,* November 20, 2021, https://www.nytimes.com/2021/11/20/world/china-congo-cobalt.html; Dionne Searcey and Eric Lipson, "Hunt for the 'Blood Diamond of Batteries' Impedes Green Energy Push," *New York Times,* November 29, 2021, https://www.nytimes.com/2021/11/29/world/congo-cobalt-albert-yuma-mulimbi.html.

7 Searcey and Lipson, "Hunt for the 'Blood Diamond of Batteries.'"

8 Searcey and Lipson, "Hunt for the 'Blood Diamond of Batteries.'"

9 Nikiforuk, "Are Electric Cars the Solution?"

10 Christy Climenhaga, "We Need Rare Earth Elements for a Greener Future, But There's a Catch", *CBC News,* August 29, 2022, https://www.cbc.ca/news/canada/edmonton/rare-earth-minerals-elements-alberta-canada-climate-change-environment-1.6558991.

11 Hannah Beech, "Can a Tiny Territory in the South Pacific Power Tesla's Ambitions?" *New York Times,* December 30, 2022, https://www.nytimes.com/2021/12/30/world/asia/tesla-batteries-nickel-new-caledonia.html.

12 Beech, "Can a Tiny Territory."

13 Beech, "Can a Tiny Territory."

14 Learn more about the Province of Ontario's Ring of Fire online at https://www.ontario.ca/page/ontarios-ring-fire. See also Emma McIntosh, "Everything You Need to Know about the Push to Mine Ontario's Ring of Fire," *The Narwhal,* August 2, 2022, https://thenarwhal.ca/ontario-ring-of-fire-explainer/.

15 McIntosh, "Everything you Need to Know."

16 Dayna Nadine Scott, "'Critical' Minerals and the Politics of Refusal,"

Toxic News, April 30, 2021, https://toxicnews.org/2021/04/30/critical-minerals-and-the-politics-of-refusal/.

17 McIntosh, "Everything you Need to Know."
18 Ian Ross, "Ring of Fire Video Draws Mixed Reaction from Women Professionals," *Northern Ontario Business,* August 16, 2016, https://www.northernontariobusiness.com/industry-news/mining-ring-of-fire-video-draws-mixed-reaction-from-women-professionals-372190. For more on the gender-based violence of colonial extractivism, see Rebecca Jane Hall's discussion in *Refracted Economies,* 2022.
19 McIntosh, "Everything you Need to Know."
20 Scott, "'Critical' Minerals and the Politics of Refusal."
21 Scott, "'Critical' Minerals and the Politics of Refusal,"
22 See, for example Cole Sayers' vision in Zoe Yunker, "The Coming Indigenous Power Play," *The Tyee,* April 20, 2022, https://thetyee.ca/News/2022/04/20/Coming-Indigenous-Power-Play/.
23 Paula Dobbyn, "Longtime Geothermal Critics Refuse to Back Down as Puna Plant Eyes Expansion," *Civil Beat,* June 15, 2023, https://www.civilbeat.org/2023/06/longtime-geothermal-critics-refuse-to-back-down-as-puna-plant-eyes-expansion/; Mark Ladao, "Controversial Kahuku Wind Project Being Connected to Electric Grid," *Star Advertiser,* August 11, 2020, https://www.staradvertiser.com/2020/08/11/breaking-news/kahuku-wind-project-being-connected-to-electric-grid/.
24 Government of Canada, *The James Bay and Northern Quebec Agreement and the Northeastern Quebec Agreement — Annual Reports 2008–2009/2009–2010,* https://www.rcaanc-cirnac.gc.ca/eng/1407867973532/1542984538197.
25 Jonas Kako, "The Fight to Save the Vjosa, One of Europe's Last Wild Rivers," *Globe and Mail,* January 30, 2022, https://www.theglobeandmail.com/canada/article-vjosa-wild-river-europe-hydroelectric-power-plant/.

PRISMATIC REFLECTIONS: CULTIVATING CARE AND COMMUNITY THROUGH MULTIFACETED CRISES

1   Christine Kelley and Michael Orsini (eds.), *Mobilizing Metaphor: Art, Culture and Disability Activism in Canada* (Vancouver: UBC Press, 2017). See also IRPP *In/Equality* podcast featuring Michael Orsini, May 5, 2023. Transcript available online: https://policyoptions.irpp.org/2023/05/inequality-and-disability-justice-transcript/.

2   Sean Craig and Carolyn Jarvis with Global News, Emma McIntosh, Sawyer Bogdan, Morgan Bocknek, and Robert Mackenzie with the Ryerson School of Journalism, "There Are Toxic Secrets in Canada's Chemical Valley," *Canada's National Observer,* October 14, 2017, https://www.nationalobserver.com/2017/10/14/news/there-are-toxic-secrets-canadas-chemical-valley.

3   Paul Morden, "Questions Being Asked Following Recent Chemical Valley Incidents," *The Observer,* February 10, 2014, https://www.theobserver.ca/2014/02/10/questions-being-asked-following-recent-chemical-valley-incidents. See also CAER (Community Awareness Emergency Response), CVECO: Chemical Valley Emergency Coordinating Organization, https://caer.ca/cveco/.

4   Details of a shelter-in-place incident elaborated by CVECO/CAER (Community Awareness Emergency Response) on their website: https://caer.ca/cveco/.

5   Clayton Thomas-Müller, "Opinion: This National Day for Truth and Reconciliation, Canada Should Commit to a Just Transition for Indigenous Peoples," *Globe and Mail,* September 30, 2021, https://www.theglobeandmail.com/opinion/article-this-national-day-for-truth-and-reconciliation-canada-should-commit-to/.

6   Christopher Flavelle, Anne Barnard, Brad Plumer, and Michael Kimmelman, "Overlapping Disasters Expose Harsh Climate Reality: The U.S. Is Not Ready," *New York Times,* September 2, 2021, https://www.nytimes.com/2021/09/02/climate/new-york-rain-floods-climate-change.html.

7 Christopher Flavelle, "Climate Change Is Bankrupting America's Small Towns," *New York Times,* September 2, 2021, https://www.nytimes.com/2021/09/02/climate/climate-towns-bankruptcy.html.

8 Christopher Flavelle, "The Cost of Insuring Expensive Waterfront Homes Is About to Skyrocket," *New York Times,* September 24, 2021, https://www.nytimes.com/2021/09/24/climate/federal-flood-insurance-cost.html. For an academic discussion of living within commodified conditions in disaster scenarios, see Jennifer Lawrence and Sarah Marie Wiebe (eds.), *Biopolitical Disaster* (London: Routledge, 2017).

9 Roxanne Egan-Elliott, "Protesters Occupy Johnson Street Bridge Tuesday Evening, before Heading Downtown," *Times Colonist,* August 31, 2021, https://www.timescolonist.com/local-news/protesters-occupy-johnson-street-bridge-tuesday-evening-before-heading-downtown-4691762.

10 Darron Kloster, "RCMP Break Fairy Creek Protester Line with Pepper Spray as Arrests Reach 740," *Times Colonist,* August 22, 2021, https://www.timescolonist.com/local-news/rcmp-break-fairy-creek-protester-line-with-pepper-spray-as-arrests-reach-740-4691556.

11 *CBC News,* "Judge Rules in Favour of Journalists' Access to Fairy Creek Blockade," July 21, 2021, https://www.cbc.ca/news/canada/british-columbia/judge-fairy-creek-media-restrictions-won-1.6112156.

12 Sarah Van Gelder and Vandana Shiva, "Earth Democracy: An Interview with Vandana Shiva," *yes! Magazine* (blog), January 1, 2003, https://www.yesmagazine.org/issue/democracy/2003/01/01/earth-democracy-an-interview-with-vandana-shiva.

13 Bruno Latour, *Parliament of Things* (website), https://theparliamentofthings.org/parliament-parlement-van-de-dingen-noordzee-ambassade-bruno-latour/.

14 Latour, *Parliament of Things.*

15 For more, read Aimée Craft's brilliant children's book, one of the first that I read to my son: *Treaty Words: For as Long as the Rivers Flow* (Toronto: Annick Press, 2021).

16 Michelle Gamage, "Tree-Sit Protesters Say TMX Is Putting Them at Risk," *The Tyee,* September 9, 2021, https://thetyee.ca/News/2021/09/09/Tree-Sit-Protesters-TMX/.

17 For more, see the Mother Tree Project website: https://mothertreeproject.org/.

18 See also Sarah Marie Wiebe, *Everyday Exposure: Indigenous Mobilization and Environmental Justice in Canada's Chemical Valley* (Vancouver: UBC Press, 2016).

19 Sierra Club BC, *Being a Voice for the Forest: A Guide to Taking Action* (website), https://sierraclub.bc.ca/action-guide/.

20 Jeremiah Rodriguez, "Gov't Disregard of Indigenous Prescribed, Cultural Burns 'Created This Catastrophe': Advocates," *CTV News,* July 27, 2021, https://www.ctvnews.ca/climate-and-environment/gov-t-disregard-of-indigenous-prescribed-cultural-burns-created-this-catastrophe-advocates-1.5525057. See also Tara K. McGee, Amy Cardinal Christianson, and First Nations Wildfire Evacuation Partnership, *First Nations Wildfire Evacuations* (Vancouver: UBC Press, 2021).

21 Audrey Carleton, "Why More Climate Change Means More Oil Spills," *Vice,* September 14, 2021, https://www.vice.com/en/article/93y4ba/why-more-climate-change-means-more-oil-spills.

22 Alex Migdal, "More than a Billion Seashore Animals May Have Cooked to Death in B.C. Heat Wave, Says UBC Researcher," *CBC News,* July 5, 2021, https://www.cbc.ca/news/canada/british-columbia/intertidal-animals-ubc-research-1.6090774.

23 Terry Tempest Williams, *When Women Were Birds: Fifty-Four Variations on Voice* (New York: Macmillan Publishers, 2012), 18.

24 Williams, *When Women Were Birds*, 20.

Index

'āina (land)
  care for, 92
abundance, 95
Aamjiwnaang, 18, 39–43, 45–47, 137, 143
  See also Chemical Valley
abundance, 17, 94–95, 129, 133, 144
  See also Beamer, Kamana; Hawai'i
academia
  care in, 49, 83
accountability, 3, 36, 140
Ahmed, Sara, 64
ahupua'a governance, 92–93
  See also Hawai'i; governance
air monitoring, 41
  See also pollution
aloha 'āina, 88, 91–92, 94–95, 128
  See also Hawai'i; policy
Aloha 'Āina Economic Futures (AAEF), 95
  See also Beamer, Kamana; Hawai'i; policy
Alook, Angele, 86, 95
Anishinabek, 49
Atleo, Richard Umeek, 54
atmospheric rivers, 30, 33, 74
Attawapiskat, 13, 15

nonviolent direct action, 142
resource extraction, 127
state of emergency declarations 11, 138
  See also Spence, Theresa

BC Coroner's report, 6
Beamer, Kamana, 88, 93–95, 109
  See also aloha 'āina; Hawai'i
Bill C-226, 19
bilobed placenta, 57, 64
birth ratio study, 45
  See also Aamjiwnaang; Chemical Valley; *The Disappearing Male*
blue bureaucracy, 17
  See also policy
blurry state of being, 16, 20, 62–63, 121, 132
bodies of knowledge, 22, 69, 135
body politic, 13–14
borders, 12, 39
breastfeeding, 32, 63, 68, 83, 102, 168n39
Brum, Eliane, 21
burden of proof, 46
Butler, Judith, 114, 116

Caesarean section, 8, 10, 22, 53, 59–61, 64, 134
capitalism
    and colonialism, 94–97
    as an economic system, 87, 90, 94, 131
    See also commodification; extractive capitalism
carbon colonialism, 98
care, 54, 133
    and love, 117–118
    and work, 50, 86, 90, 147
    based social infrastructure, 85
    beyond extraction, 36, 80, 99, 101, 138, 146
    beyond individualism, 11, 25
    counter to capitalism, 18, 21, 31, 94, 128
    counter to crisis, 2, 4, 23, 31
    crisis of, 89–90
    See also Fraser, Nancy; social reproductive labour)
    ethos of, 25, 47, 50, 119, 146
    for children, 62, 85, 133, 147
    mutual, 81
    networks of, 24, 142, 147
    politics of, 158n26
    See also healthcare
care-based economies, 18, 21, 86–87, 100–101, 128
ceremonial fast, 15
    See also Attawapiskat; Spence, Theresa
Chemical Valley, 16, 18, 39–41, 46–47, 49, 110, 137
    See also Aamjiwnaang
circular economy, 18, 94, 130, 132, 147
    approaches, 88–89, 91
    of care, 85, 97
    teaching about, 96
climate coloniality, 76
Climate Change Conference of the Parties (COP), 34, 97–98
climate displacement, 33
coastlines, 72–73
    and feeling, 76, 145
    See also flooding
climate emergency
    and care, 21, 86, 98
    and gender, 28, 97
    and layered crises, 22, 27, 53–54, 67, 81, 90, 135, 146
    as public health emergency, 35, 134–135
    as twin crisis of climate and democracy, 132
    causes of, 79, 91
    experiential knowledge of, 29, 78, 132
    feelings, 9, 137
    lived experience, 69, 132, 136, 142
    politics of, 9, 31
    solutions to, 81, 91, 101, 122, 129, 132
    reprieve from, 29
    urgency, 20, 25, 83
climate grief, 11, 82–83
Coast Salish territory, 55, 78, 95–96, 139, 142–144, 148
co-creation, 50, 129, 150

co-design, 96
collective action, 87, 98, 114, 117, 138, 141
    Black Lives Matter, 19
    Fairy Creek, 139–140
colonialism
    and extraction, 47, 76, 112, 125
    and capitalism, 94–97
    and military, 107–108
    and state power, 3, 130
    legacy of, 15, 40, 70
colostrum, 62
commodification, 87–88, 92, 131
    See also capitalism; transactional relations
community
    advocacy, 41, 45, 95, 99, 114, 141
    affected by layered crises, 13, 15, 19, 33–34, 39, 42, 73, 75, 112, 134, 143, 145
    and care, 19, 21–24, 83, 86, 88, 101, 116–117, 119, 134, 147, 150
    and race, 19, 35, 76, 99, 110, 112–113
    and relationships, 24, 133, 148, 44
    and states of emergency, 15, 114, 138
    beyond extraction, 115, 126
    beyond individualism, 116
    building, 90
    ceremonies, 81
    coastal, 34, 69, 78–80, 95
    consultation, 127
    displacement, 73, 80
    diversity, 117
    economics, 139
    education, 50, 96
    face of environmental protection, 41
    frontline, 97–98
    gardens, 147
    governance, 94
    health and well-being, 7, 11, 19, 21, 25, 41, 48, 74, 81, 91, 113, 124, 141, 159n1
    in Hawai'i, 78, 93–95, 130
    inequity, 76
    knowledge, 99
    leadership, 48, 108
    led-solutions, 98
    low-income, 46
    marginalized, 112
    networks, 23, 142
    of colour, 35, 76, 112
    political communities, 14, 42, 116
    politics of, 22, 46
    power of, 118
    rebuilding, 6
    resilience, 139
    sense of, 64, 118
    stories, 50, 81
    sustainable development of, 92
    violence, 112–113
    visions, 96, 129
    voices, 43, 129
community-based research, 44–45, 48
community engaged-research, 48–51, 142, 159n
consent, 22, 59, 130, 138

converging crises, 28, 31
Cook, Katsi, 55
COVID-19, 12, 27, 31, 89, 100, 113
   Building Bridges, Not Walking on Backs: A Feminist Economic Recovery Plan for COVID-19, 31, 100
corporeal citizenship, 103
   See also Gabrielson, Teena; Parady, Katelyn
counternarratives, 22
crisis
   and states of emergency, 10
   converging or layered, 22, 27–28, 31, 54, 67–68, 74, 82, 90, 133–135
   experiences and experiential knowledge, 69
   felt through the body, 9
   health care, 9, 135
   housing, 15, 135
   in relation to water, 76
   mental health, 11, 13
   multidimensional, 90
   of democracy, 132
   planetary health, 11
   politics of, 10
   racialized, 113
   stories, 69
   sudden, 100
   wildfire, 16
critical disabilities studies, 135
critical policy studies, 12
cumulative impacts, 45
   See also pollution

De Beers Victor Diamond Mine, 126
decision-making
   and authority, 29, 36, 140
   and policy, 116
   power, 113, 132
   State of Emergency declarations, 12
Declaration on the Rights of Indigenous Peoples Act (DRIPA), 128
decolonial futures, 22, 92, 96–97, 108, 128–129, 138
decolonial thought, 11, 87, 108
democracy
   and climate change, 132
   and policy, 114
   and public engagement, 9
   deliberation, 117
   disregard of, 113, 124
   earth, 140, 144
   fleshy, 103
   health of, 115
   life, 13, 73, 114
   practices and norms, 12
   processes, 108
   transparency and accountability, 140
   underpinnings and foundations, 113, 115
Descartes, René, 51
design, 43
Dickinson, Emily, 61
disaster, 10, 16, 20, 39, 47, 126, 139
   biopolitical, 47
   See also mass casualties
discourse, 12, 19, 36, 45, 115

divestment, 87, 99
documentary film, 45, 49, 53
   Indian Givers, 49
   See also The Disappearing Male
doula, 58
drought, 7, 30
dualisms, 14, 51, 97, 101, 142

earth democracy, 144
   See also Shiva, Vandana
ecofascism, 110
ecofeminism, 11, 86, 88, 97, 140, 142, 158n26
   definitions of, 22, 101–103
economics, 47, 86, 88–89, 95, 97–99, 127–128
EJAtlas (Global Atlas of Environmental Justice), 100
   See also environmental justice
electric vehicles, 20, 122–123, 125, 128
embodiment, 51, 93, 102, 114, 133, 136
empire, 39, 108
   See also colonialism
emotions, 44, 51, 54, 63–64, 90, 134
   See also feelings
energy
   clean, 122, 127, 129, 131
   consumption of, 128, 132
   extraction, 125
   futures, 128, 131
   renewable, 20, 99, 127–128, 138
   revolution, 121
   sector, 20, 127, 129
   tidal, 129
   transition, 122, 126

environmental justice
   activism, 92, 142
   and extractivism, 98
   conflicts, 20
   scholarship, 11, 46, 99, 131, 142
   uneven impacts of, 96
   See also Bill C-226; EJAtlas
environmental injustice, 13
Environmental Noxiousness, Racial Inequities and Community Health (ENRICH) project, 19, 113
   See also Waldron, Ingrid
environmental racism, 19, 113
   See also Bill C-226; Waldron, Ingrid
environmental violence, 18, 113
   See also slow violence
equity, 12
evacuation, 18, 27, 137–138
executive power, 9, 36
   authoritative decision, 12
   head of state, 15, 36
experiential learning, 96
extraction
   and colonialism, 3, 112
   and consent, 138
   and sacrifice zones, 115
   and war, 100
   dangerous logic of, 127
   fossil fuel, 99
   from nature, 69
   of critical minerals, 20, 122, 125–126, 128
   of knowledge, 50
   of resources, 21

modes of, 20, 121
renewable, 121
sector, 100
sites of, 80, 124
extractive capitalism, 11, 18, 21, 47, 87, 92, 112
in policy, 98, 100
in research, 50
See also capitalism

feelings, 48
about emergencies, 14, 113
about multilayered crises, 135
flesh and, 6
non-Western, 81
of isolation, 29
related to birth, 149
See also emotions
felt theory, 22
feminism, 87–88, 97, 100–101
See also ecofeminism
Feminist Environmental Research Network (FERN), 23–24, 148–149
Feminist Green New Deal Coalition, 97–99
feminist parenting, 23–25, 89
fire
and feminist mothering, 27
and pregnancy, 27, 29, 143, 148
and states of emergency, 11, 144
BC wildfires, 6–7, 10, 33, 144
Code 9, 137
control and management, 145
disaster of, 47
element of, 143

exposure to, 27
impacts of, 28, 67, 144
fleeing, 10
flooding and landslides, 74, 137
global threat of, 16
Indigenous approaches to, 143–144
Lahaina, Maui, 78
related to climate change, 30, 68
smoke, 27–28, 143, 148
spread of, 11, 30
threat of, 16, 53, 55
flesh, 9, 34
fleshier democracy, 103
flooding
and climate change, 7, 30, 67–68, 72
and planetary health, 145
flash, 77
in BC, 35, 73, 137
in Hawai'i, 77
in New York, 139
insurance, 139
slow disaster of, 47
See also coastlines
Fraser, Nancy, 31, 87, 89–90
See also care, crisis of; social reproductive labour
freedom convoy, 113, 115
friendship, 118, 147
futures
ʻĀina Aloha Economic Futures (AAEF), 95
alternative, 5, 37, 89, 96, 101, 108, 128–129, 132, 136, 138, 140
care-centred, 128

climate, 36
concerns with, 29
decolonial, 22, 92, 108, 128–129, 138
ecological, 132, 140
economic, 89, 95–96
energy, 128, 131
environmentally just, 135
feminist, 88
healthy, 13, 31, 37, 88
post-extractive, 99
radical policy, 116
reproductive, 39, 47
sustainable, 5, 31, 35, 88, 96, 128–129, 131, 138

Gabrielson, Teena, 101, 103
gender, 28, 36, 111, 127
George, Rachel yacaaʔał, 1–5, 92, 148
governance, 12, 17, 46, 49, 130, 139
community, 94
ecological, 93–94
See also blue bureaucracy; state of emergency
green consumerism, 89, 92, 122, 124
Green New Deal, The, 96
See also Red Deal, The
green theory, 101
grief, 11–12, 14, 17
Guterres, António, 34
See also United Nations

Hall, Rebecca Jane, 99
Harper, Stephen, 15
Hawaiʻi
ahupuaʻa, 79, 92–93
and climate change, 78
and colonization, 107–108
ecological governance, 94
emergency alerts, 12, 77, 105–107
Emergency Management Agency, 77, 105
environmental justice, 75
flooding, 74
geopolitics, 106
geothermal developments, 130
housing, 12
Lahaina fires, 78
leadership, 11, 77
loʻi, 92
Maui, 11, 78, 144
Mauna Kea, 12
militarization, 108–109
occupation, 107–108
reports, 31
researchers, 56, 78, 93, 107
State Commission on the Status of Women, 31, 100
State of Emergency declarations, 12, 77
storms, 74, 76–77
See also ahupuaʻa governance; *Building Bridges, Not Walking on Backs: A Feminist Economic Recovery Plan for COVID-19*
health authorities, 10
Bluewater Health (Lambton County), 39, 137
Interior Health (Kelowna), 10
healthcare, 21, 25, 31, 59, 83, 94

access to, 9, 135
emergency, 8, 11, 35, 134
primary, 85, 134
heat, 7, 11, 27–28, 31, 35, 54, 137, 144, 146
heat dome, 1, 6–8, 30, 33, 35, 145, 149
hierarchical relations, 9, 17
hooks, bell, 19, 25, 116–117
hospital codes, 15–18, 20, 47, 137
as metaphor, 11, 15, 131, 135–138
hospitals
BC Women's Hospital, 27
Kelowna General Hospital, 10
Royal Jubilee Hospital, 10
Victoria General Hospital, 57
visits to, 8, 56–58, 62, 134
housing, 11, 15
humility, 142
hydroelectricity, 122
James Bay, 130
Vjosa River, 130–131

Imperial Oil, 137
Indigenous
communities, 86, 95, 97, 113, 144–145
consent, 126, 128, 130
economies, 18, 87
governance, 16, 53
laws, 46
leadership, 130
peoples, 3, 36, 109, 139, 141, 144
political thought, 11, 142
sovereignty, 13, 40, 128–129, 132, 138, 148
ways of knowing, 54
See also decolonial thought; treaties
Idle No More, 11, 15
individualism, 18, 21, 23, 25, 114
and lifestyle choices, 79, 102, 117
anthropocentric, 114
atomistic, 147
beyond, 3, 11, 96, 103, 116, 118–119, 147
management, 79, 91
rationalism and, 102
infrastructure
affecting planetary health, 136
breakdown, 137
built, 81, 121, 126
care, 91
cooling, 7
damage to, 74, 75, 77
extractive, 71, 141
green, 20, 73, 122, 125–126
imposition of, 131
social, 31, 85, 87, 91, 129
transportation, 129
Intergovernmental Panel on Climate Change (IPCC), 34, 75
intersectionality, 25, 28, 97
involution, 63

James Bay
Cree, 130
peatlands, 126
See also hydroelectric power
James Bay and Northern Québec Agreement, 130

Johnston, David, 15
joy, 61, 133, 146–147
jurisdiction, 93, 112, 114, 138
justice, 19, 86, 88, 97, 116

kanaka ʻōiwii, 88
    See also Hawaiʻi
kinship, 3, 9, 99, 118, 142, 147
    Lambton Community Health
        Study, 41, 48
    See also Aamjiwnaang;
        Chemical Valley
Land Back, 95
Latour, Bruno, 140
    See also parliament of vital
        things
listening
    and imagination, 23, 129
    and more-than-human environ-
        ments/relations, 142
    and transformation, 129, 144
Lockridge, Ada, 41, 43–44, 47
    See also Aamjiwnaang
love, 117–119
    and pain, 16, 40, 54, 61, 63–64
    See also yaʔakmis
Luke, Tim, 9
Lytton, BC, 6, 33, 35

mass casualties, 10, 47
    See also disaster
McGregor, Davianna Pōmaikaʻi, 93
    See also ahupuaʻa; Hawaiʻi
mental health, 7, 11–12, 46, 82
    prenatal depression, 29
    postpartum, 17, 31, 54, 67, 149
    post-traumatic stress disorders, 27
    reproductive, 22, 61
    suicide, 112
mesothelioma, 41
    See also Chemical Valley
midwives, 57–58
Million, Dian, 22
militarization, 92
    and environmental impacts, 80,
        100, 110, 123
    in the Democratic Republic of
        Congo, 123
    health impacts of, 108, 110
    in Hawaiʻi, 95, 106–109
Mother Earth, 23, 95, 144
mothering, 16, 27
    a blurry state of being, 63, 121, 132
    feelings and experiences of, 8,
        31, 81
    and body, 32, 101
    See also feminist parenting

natural resources
    and capitalism, 17, 87–88, 131
    beyond profit, 127
    peatlands, 126
    value of, 94
    See also energy; environ-
        mental justice; extraction;
        Indigenous; sacrifice zones
Nixon, Rob, 16, 46
    See also slow violence
non-violence, 114–116, 140, 142
    See also Butler, Judith

Northern Ontario, 13
    See also De Beers Victor Diamond Mine; Ring of Fire
nursing (feeding infants), 32, 63, 68, 83, 102, 168n39
Nuu-chah-nulth, 16, 54, 139
    See also yaʔakmis

Oʻahu Water Protectors, 108
    See also Hawaiʻi
oceans
    acidification, 145
    and mothering, 2, 7, 145–146
    and wellness, 17, 68
    as harbingers of planetary health, 68
    as life force, 69–71, 92
    as metaphor, 17, 145–146
    erosion, 30, 72
    glacial meltwaters, 72, 145
    mass death of ocean species, 68
    National Oceanic and Atmospheric Administration study, 72
    governance of, 17, 69, 92–94
    oceanographers, 74
    Pacific Ocean, 69–71, 74, 77, 78, 80, 92, 106, 109, 111, 125, 145
    submarine volcanos and tsunami, 71–72
    warming of, 72, 74
    See also Pacific Ocean; sea-level rise; seascapes; water
Oceania, 79–80
Okanagan, 144

Orsini, Michael, 12
Osorio, Heoli, 107

Pacheedaht, 130
Pacific Ocean, 69–71, 74, 77–78, 80, 92, 106, 109, 111, 125, 145
Parady, Katelyn, 101, 103
parenting, *See* feminist parenting
parliament of vital things, 140–141
    See also Latour, Bruno
patriarchy, 87, 95
planetary health, 7, 22, 68–69, 90, 145
    and transformative change, 45, 136, 166n14
    beyond extractive capitalism, 11, 18, 133
    in relation to Western ways of knowing, 21, 92
    personal impacts of, 10–11, 15, 82
    political implications of, 10, 15, 22, 82, 122
    voice and speaking out about, 140
poetry, 108
police violence, 112–113
    See also collective action
policy
    administration, 69
    alternative configurations of, 68
    and democracy, 9
    application of, 94
    assemblage, 12, 43, 46
    care, 101
    climate change, 17, 97–100, 121
    conversations, 25
    critical policy studies, 12, 79, 135

debates, 36
decisions, 42, 69, 113
experts/expertise, 129
extraction, 100
feminist, 97–100
implications, 86
in Hawai'i, 95
Indigenous, 48
inequity, 19
instruments, 13
justice, 135
multifaceted, 135
multijurisdictional, 42. 128–129
outcomes, 69, 114
radical, 101, 116
reports, 138
solutions, 94, 98
transformative, 29, 129, 139
violence, 19, 111
See also blue bureaucracy; aloha 'āina
pollution, 47, 49, 108–109, 137
health concerns caused by, 27, 76, 110, 143
See also Chemical Valley
portal,
birth as, 118
body as, 61, 64, 118, 146
emergencies as, 101
to imaginative thought about alternative realities, 17, 101
water as, 76
power
and domination, 117
and emergencies, 135
and governance, 20
and violence, 18
asymmetrical power relations, 13
decision-making, 113
extractive, 118
hierarchical, 9
in community and collective struggle, 114, 118, 138
Indigenous perspectives of, 2, 128
of arts, 81
of oceans, 146
of storytelling, 81
outages, 77–78
political, 108
relations, 46, 131
pregnancy, 1–2, 28–29, 32, 53–56, 68, 82, 139, 146, 148
public engagement, 9, 142
consultation, 12
public health, 12, 35
See also COVID-19; healthcare; planetary health

racism, 25, 111, 112
environmental racism, 18–19, 99–100, 110–114
See also Black Lives Matter; ENRICH project; police violence; Waldron, Ingrid
reciprocity, 3, 11, 22, 50, 87, 102, 133, 136, 142
Red Deal, The, 96
See also Green New Deal, The
Red Hill storage facilities, 108–109
See also Hawai'i; militarization

regeneration, 18, 20
relationships
    and disasters, 16
    and planetary health, 20, 117, 133, 144
    caring, 23, 25, 99, 101, 138, 147
    extractive, 4, 23, 50
    feminist, 24
    healthy, 22, 24, 37
    Indigenous teachings about, 4, 50, 71, 99, 144
    intimate, 94
    research, 16, 44, 50, 143
    See also kin
reproductive justice
    Aamjiwnaang, 39, 45, 47–48
    and feminist parenting, 25
    and health futures, 39
    impacts of fossil fuels and petrochemicals, 98, 102
renewable energy, 20, 99, 127–129, 138
renewable renaissance, 122
resurgence, 49
Ring of Fire, 126–127, 138, 173n14

sacrifice zones, 115
Salish Sea, 141
Sarnia, ON, 39, 48–49
scars, 64
    as metaphor, 65
    Caesarean section, 61, 64
    See also Ahmed, Sara
sea-level rise, 30, 33–34, 72
sea life, 17
    See also oceans; seascapes; water

seascapes, 16, 92, 154n27
    See also oceans; sea life; water
sensory experiences, 28, 42, 62, 138
    sight, 29, 42
    sound, 29, 43
    smell, 29, 42–43, 46
    taste, 29, 42, 63
    touch, 29
    See also scars
shelter-in-place, 43
    See also Aamjiwnaang; Chemical Valley
Shiva, Vandana, 140, 144
    See also earth democracy; ecofeminism
Sierra Club BC, 30, 144
Simard, Suzanne, 142
slow crises, 33, 138
slow violence, 15–16, 46–47, 105, 112, 138
    See also environmental violence; Nixon, Rob
social reproduction, 90, 99
solidarity, 19, 22, 50, 114
sovereignty, *See* Indigenous sovereignty
Smith, Dawn, 16, 53–54, 148
Spence, Theresa, 15
    See also Attawapiskat; ceremonial fast
Spinoza, Baruch, 51
social reproductive labour, 86–87, 89–90, 99
    See also care, crisis of; Fraser, Nancy

state of emergency
	and climate change, 4
	and democracy, 13, 132
	as embodied event, 6
	Attawapiskat, 11, 138
	British Columbia, 10, 33, 35, 73–74, 144
	declarations, 9
	experiences of, 13–14, 83
	Hawai'i, 11–12, 77
	politics of, 9–13, 132
	research about, 138
storytelling, 22, 29, 65, 135
shorelines, 3
sustainable futures, 31, 35, 88, 128–129, 131, 136, 138–139
sustainability 18, 121
	and care, 4, 136
	and decolonization, 96
	and ecological governance, 93
	beyond capitalism, 97, 128
	economics of, 88, 91–92, 96, 122, 124, 130
	in Hawai'i, 92
	rhetoric, 18, 92, 123, 126
sweat ceremony, 48, 81

Tempest Williams, Terry, 145–146
The Climate Baby Dilemma, 53
	See also documentary film
The Disappearing Male, 45
	See also documentary film
Thomas-Müller, Clayton, 138
Thompson, Debra, 112

toxic exposure, 43, 45–46, 110
	See also Chemical Valley; pollution
toxicity, 125
transactional relations, 4, 17, 87–88, 92
transformative change
	and concrete action, 138
	and leadership, 97
	and mothering, 149
	and the environment, 45, 143
	beyond extractivism, 86, 89
	deliberations about, 100
	learning from embodied knowledges, 69
	of economic and political systems, 91, 95, 99, 130
	through place-based education, 96
	through policy, 29
	with care at the centre, 86, 95, 146–147
Trans Mountain Pipeline, 71, 141
trauma, 10, 15, 17, 53, 61–62, 133–134, 149
treaties, 12, 15, 46
	See also Indigenous; James Bay and Northern Québec Agreement
Treaty 9, 126–127
Trump, Donald, 12, 107

United Nations, 16, 23–24, 122, 128
	See also Climate Change Conference of the Parties

violence, 110–112, 115
    See also environmental violence; non-violence; police violence; slow violence
voice, 31, 36–37, 129, 132, 135, 141
vulnerability, 7, 27, 107, 139

Waldron, Ingrid, 18, 99, 110, 113, 131
water
    access to, 21
    and extraction, 124
    and human sustenance, 102, 143
    and trees, 142
    birth and, 2, 56, 58, 145–146
    bodies of, 12, 17, 68, 69–70, 82
    contamination, 13, 18, 108–110, 124
    cooling elements of, 7
    crises, 76, 78
    governance, 68, 93–94
    healing elements of, 2, 146
    hydration and dehydration, 8–9, 32–33, 82
    Indigenous relations to, 81
    in Hawai'i, 92–94, 108, 110
    is life, 95, 109
    our bodies are made of, 101, 145
    protection of, 30, 92, 108
    vapour, 74
Western knowledge, 21, 51, 101, 117
white blood cells, 60, 63
witnessing, 144
Women's Earth & Climate Action Network (WECAN), 97, 99
    See also ecofeminism; feminism

workers' rights, 28
W̱SÁNEĆ, 144, 148

yaʔakmis, 16, 54, 161n2
    See also love and pain; Nuu-chah-nulth
Yellowknife, 11, 144

Introduction

    About Me 4

    What Is Sky Wisdom? 5

    What Is Ati Yoga? 6

    Discovering Your True Nature through Sky-gazing 6

    Identifying with the Sky 7

    How to Use the Breath 8

    About the Themes 8

    Using the Sky Wisdom Oracle Cards 10

    How to Consult the Oracle Cards 11

    Using the Mantras 11

    What Are Pith Instructions? 12

    The Key to Sky Wisdom Meditation 13

    What Are the Benefits of Sky Meditation? 14

    A Final Word ... 15

The Cards

    Freedom 17

    Clarity 31

    Spaciousness 45

    Essence 59

Introduction

Welcome to the Sky Wisdom Oracle deck!

Based on the wisdom tradition of the Himalayas, the 48 cards feature photos of the sky in all its forms. Along with the accompanying guidebook, the cards are designed to inspire you to meditate with the sky to encourage relaxation and openness, helping you to let go of everyday stresses so that you may find inner peace and experience yourself in your most expanded form. The deck is loosely based on Tibetan Buddhist teachings that remind us how to connect with our true nature and become one with the infinite.

About Me

I am the founder of The Way of Meditation and have been teaching meditation since 2003. My aim is to teach authentic powerful meditation techniques that are easy to understand and practise in the modern world.

I spent six years as a Buddhist monk, living in a monastery in Queensland, Australia, studying and practising Buddhist meditation. During that time, I had deep-awakening experiences in my isolated hut in the forest and also discovered the mystic teachings of sky wisdom contained within the Dzogchen and Mahamudra traditions.

This highest level of philosophical wisdom found in Tibetan Buddhism was in alignment with my experiences and seemed to be the ultimate spiritual truth everyone is searching for. However, I was told that these teachings were not meant for beginners. I found this hard to accept, so I decided to break with my religious tradition and start teaching sky wisdom to anyone who was interested. And it turns out there are plenty of people who wish to learn about these profound practices. After my in-depth study and experience of Tibetan Buddhism, I left the monastery and explored many other mystic traditions. I discovered that sky wisdom was contained in them all. Known as the perennial philosophy, I found sky wisdom meditations in Zen, Advaita, Shaivism and Tantra, among others. I now teach these practices and spread their profound message of oneness and wholeness in meditation retreats in Australia, and my mission is to empower as many people as possible to remember their divine source and innate perfection.

What Is Sky Wisdom?

According to the Dalai Lama, the sky is a metaphor for the immense and spacious nature of the mind. Just as the sky is limitless and free of boundaries, so too is the mind, which is not limited by thoughts, emotions or perceptions but has the potential for boundless clarity and wisdom.

Drawing its inspiration from Tibetan Ati Yoga (see page 6) and the practice of sky-gazing meditation, sky wisdom is the wisdom of emptiness – of an open mind connected with vast space and infinite possibilities. It is seeing that the true nature of reality is beyond words and concepts. Being able to simply rest in this pure natural state of awareness is the **meditation of sky wisdom**.

The sky itself is the union of space and light and so the sky nature of your mind is the union of space and awareness. You can see why the sky is one of the best metaphors for your true self. Directly knowing the essential nature of yourself is pure, unbounded awareness. This is the **view of sky wisdom.**

A sense of unity with all humans flows naturally from experiencing such awareness. As we realize we all have a common core, we become more loving and can practise "loving kindness" as a way of embodying the qualities of the sky in our own lives. This simply involves behaving as if all people and the Earth itself are all intimately interconnected and united. This is the **loving attitude of sky wisdom.**

What Is Ati Yoga?

Sky wisdom and sky nature are the core principles of the highest Tibetan Buddhist wisdom tradition known as Ati yoga or the union with the ultimate, which is at the heart of the philosophy of all the great mystic traditions. Mystics all point to this true essence as being a universal field of sky-like oneness connecting all people and all things.

The word "Ati" means "the unity of pure awareness and primordial purity". Our inner nature is pure like the sky because it can never be stained or harmed by passing conditions, just as the sky above is never harmed or stained by the weather.

Ati yoga is my inspiration for these oracle cards, guided by the Buddhist teachings of the great 14th-century Ati yogi, Longchenpa (1308–64).

The Dalai Lama says that meditation on the sky through Ati yoga is the most supreme and fastest path to enlightenment (becoming one with the infinite) because of its direct approach.

Discovering Your True Nature through Sky-gazing

The sky is a symbol that represents the ultimate nature of reality, which is vast, limitless and "all-pervading". Contemplating the sky is a powerful tool for meditation because it helps you to connect with this limitless nature of reality and realize your own true nature, which is also vast, limitless and all-pervading.

By meditating with the sky as your inspiration, you allow your mind to rest in its natural state, which is spacious, open and clear, like the sky above, allowing everything to be as it is, without judgement or resistance.

A key aspect of this meditation is using the sky to realize that just as clouds and weather do not harm the sky, your thoughts and emotion cannot harm the nature of your mind. It's this indestructible sky nature that we are trying to connect with. Or more accurately, re-connect with, because our sky nature already exists as the basis of our every experience.

The practice of sky-gazing is grounded in the understanding that the nature of our mind is fundamentally pure and free from any inherent flaws or limitations – it is already perfect and complete like the sky. And just as the sky above is not obstructed by any object that appears within it, so our true nature is not limited by concepts or thoughts that appear within it.

When we rest our mind in an authentic natural state through meditation, we can access a profound level of awareness that is beyond the ordinary mind and the limitations of dualistic thinking (black/white, good/bad, negative/positive and so on).

Sky wisdom gives us a portal into the everywhere-all-at-once experience. From this, unity with all humans flows naturally and we realize that we have a common core, allowing love and empathy for all to flourish.

Identifying with the Sky

The sky is a universal symbol in all cultures and traditions, and it is therefore a powerful archetypal image that can resonate deeply within all human beings around the world. It is a pure and unchanging presence that is always there, no matter what is happening on the surface of the Earth.

This can serve as a powerful reminder to us that our true nature is also pure and unchanging, and is always present, even when obscured by thoughts, emotions and other distractions.

The vastness and openness of the sky can help us to let go of our habitual patterns of thinking and feeling and open up to a more expansive and inclusive way of being. I encourage you to identify more deeply with the sky, so that you begin to realize your identity is not limited to your physical body or your personality but is, in fact, a part of the vast and limitless nature of reality itself.

It's the immediacy and availability of our ultimate true nature that sets sky meditation practices apart from other, more gradual approaches to meditation.

Through the practice of meditating on the sky, we can intuitively recognize the natural state of our mind and rest in this experience. This direct approach is said to be the fastest path to enlightenment because it goes straight to the enlightened truth of every experience.

How to Use the Breath

The cards often ask you to pause and take a breath. It's best to always breathe through your nose if you're able to. This helps slow the breath down, filters the air and even activates hormonal chemicals that boost the immune system. It's important to note that the breathwork exercises in the deck are intended to help calm you down and therefore long, slow, deep breaths are recommended. These help to quieten the nervous system and make it easier for you to notice the subtle inner world of sky wisdom.

About the Themes

The 48 Sky Wisdom Oracle cards are arranged in four themes of 12 cards, which address the same wisdom from slightly different perspectives.

THE FOUR THEMES
- Freedom
- Clarity
- Spaciousness
- Essence

FREEDOM
"True freedom is achieved when we are able to see through the illusion of our own projections." LONGCHENPA

The sky is vast and boundless, without limitation or constraint. Similarly, freedom lies within us and is always accessible, but we often create limitations and constraints through our beliefs and attachments. The sky does not cling to the clouds that pass through it, nor does it resist them. In the same way, we can learn to let go of our attachments and allow experiences to pass through us without resistance or grasping. True freedom comes from accepting and embracing the impermanence of all things.

CLARITY
"Clarity is the natural state of the mind, obscured only by our conceptualizations and habitual patterns of thinking." LONGCHENPA

The sky is clear and transparent, reflecting all that is in it. In the same way, we can cultivate clarity by allowing our thoughts and emotions to arise and pass without getting caught up in them. The sky does not judge or analyse what it reflects; it simply is. Similarly, we can learn to observe our experiences without judgement or analysis, allowing ourselves to see things as they are. This clarity helps us to make decisions from a place of insight and wisdom.

SPACIOUSNESS
"The spaciousness of our being is not something that can be created or destroyed; it is always present, waiting to be realized." LONGCHENPA

The sky is spacious, with room for everything that exists. When we cultivate spaciousness within ourselves, we open up to new possibilities and experiences. When we let go of limiting beliefs and assumptions we create space for growth and transformation. The sky does not discriminate or hold back its vastness; it is always available to connect with. In the same way, when we cultivate spaciousness, we connect with a sense of openness and possibility that allows us to live more fully.

ESSENCE
"The essence of our being is pure awareness, unbound and unchanging." LONGCHENPA

The sky has no fixed form, colour or substance; it is empty yet present in all things. Our essence is like the sky, formless yet present in everything we do. We often get caught up in the form of things – the labels and identities we attach to ourselves and others. But like the sky, our essence is always there, beyond labels and forms. When we connect with our essence, we tap into deeper wisdom and compassion that go beyond our limited understanding of ourselves.

Using the Sky Wisdom Oracle Cards
The Sky Wisdom Oracle cards can be used for **daily meditation and inspiration** to aid you in remembering your vast and sky-like nature, and to help you to move beyond the narrow confines of ego and limitations of dualistic (black/white) thinking.

Each card is a reminder of your essential nature of limitless potential and creativity, bringing space and light for you to connect with in every moment so you can overcome stress and negative thinking and **reach your full potential**.

Learning to recognize and rest in your sky nature is **enlightenment itself.** When you rest your mind in this natural state, you **gain access to a profound level of awareness** that is beyond the ordinary mind.

How to Consult the Oracle Cards

Shuffle the deck and **allow your intuition** to choose a card. Imagine the card is perfectly synchronized with your spiritual journey and will give you the reminder that you personally need the most at that time you choose to draw it.

Each card features a key word – use this as a simple prompt to help you unlock the meaning of the card and to guide your insight and understanding. Think of it as a personal instruction meant just for you, so that you can contemplate and integrate its theme into your life. When you have drawn your card, look it up in the guidebook. Allow the accompanying **aspiration, reflection, meditation** and **mantra** to stay with you for the day, the week or until your chosen card's theme has revealed an enlightened aspect of yourself. Be patient using the cards and let the insights come naturally through using them often.

Contemplating the guidance alongside the cards regularly will bring you a limitless dimension of wisdom, stillness and awakening to integrate into every aspect of your life.

Using the Mantras

As mentioned, each card entry includes a mantra. These mantras are positive statements, which, when repeated, often aid contemplation.

By seeing the qualities or attributes they refer to as ones you already possess, you can rediscover lost dimensions of yourself. This will help you to develop self-awareness of aspects of yourself you may have forgotten about.

Also, framing a positive statement in the present tense – for example, "My mind is clear and free" – is a powerful psychological tool to help you embody the qualities and attributes it mentions. Just repeating the mantra to yourself can have positive effects on your mind and improve the quality of your day.

What Are Pith Instructions?

The Sky Wisdom Oracle cards follow the Himalayan wisdom tradition of teaching enlightenment through **pith instructions**, which consist of concise, short teachings that help us to penetrate our understanding of ourselves and reach enlightenment more directly.

These oracle cards contain the pith instructions of sky wisdom, the highest approach to meditation, presented in the **aspirations, reflections, meditations** and **mantras.** They are all designed to be easy to contemplate and recall.

Remember, the ultimate nature of reality is already present within you, but it is obscured by your habitual patterns of thought and behaviour. Each of these cards can help to unveil your hidden brilliance and reveal your inherent good qualities.

By contemplating and integrating the teachings of these sky wisdom pith instructions, you can begin to unravel negative patterns and traumas, and access your innate wisdom and clarity. Through this process, you can gradually internalize the instructions and start to embody their essence in your daily life. Always take these cards as personal advice, just for you to contemplate and integrate into your own life.

Each card is named after a word connected with sky wisdom and has its own aspiration, reflection, meditation and mantra. Let the **aspirations**, which are designed to orient your mind toward the wisdom encapsulated in the cards, help you realize the innate qualities and attributes they promote. Allow the **reflections** to lead you into an open expansive field within, to directly experience the clarity and spaciousness these cards point to. Use the **meditations** to practise and experience your sky nature directly. It can be especially powerful to use the cards as personal meditation instructions. And let the **mantras** be your daily prayers, repeating them throughout the day many times to connect with the messages within.

The Key to Sky Wisdom Meditation

Sky wisdom meditation emphasizes the natural authentic state of being, the eternal pure primordial awareness at the heart of all beings. This Ati yoga meditation style adopts an easy-going, non-striving and non-doing approach to meditation practice, similar to Zen.

The fundamental principle of sky wisdom is the recognition of the innate awareness that pervades all phenomena. As this awareness is already perfectly whole and complete, it is not something that needs to be achieved through meditation or any other means – it is already our authentic nature. It is already present within us, but we are not aware of it because we are distracted by our thoughts and emotions. The practice of sky wisdom meditation is therefore not about doing something to attain this awareness but rather about recognizing and resting within it. And it also entails clearing away the clouds of confusion and distraction to reveal the totally free and unbound nature that is eternally present within us.

There is no need for complicated meditation techniques or formal practices. Instead, the focus is on relaxing and letting go of any effort to achieve or attain anything. Simply rest in the natural state of being. You don't need to strive or try to manipulate anything. Just let go of any effort or intention and allow yourself to be in the present moment, free from any distractions. The cards will continually remind you of this.

Each card will help to clear away the clouds of confusion to reveal your perfectly existing sky wisdom. Meditate without seeking any particular goal or outcome. True realization will arise spontaneously, without any effort or striving on your part. Simply sit and contemplate the sky without any agenda or goal in mind and allow your true nature to reveal itself naturally.

What Are the Benefits of Sky Meditation?

Meditating and contemplating the sky is a powerful practice for gaining insight into the nature of reality and realizing your own true nature. This practice will help you to:

1. CULTIVATE SPACIOUSNESS
By meditating on the vast and open expanse of the sky, you can learn to cultivate a sense of spaciousness and openness in your mind. This can help to loosen the grip of habitual patterns of thinking and feeling and create more room for insight and clarity to arise.

2. OVERCOME ATTACHMENT
As the sky is a symbol of boundlessness and freedom, meditating on the sky can help you to let go of your attachments and aversions. By contemplating the sky, you can learn to cultivate a non-grasping, non-clinging attitude toward your experiences and emotions.

3. CONNECT WITH THE ULTIMATE NATURE OF REALITY

The sky is also a symbol for the ultimate nature of reality, which is vast, unchanging and all-pervading. By meditating on the sky, you can connect with this ultimate reality and deepen your understanding of the true nature of yourself and the world.

4. ENHANCE AWARENESS

By focusing on the sky and its qualities, such as vastness, clarity and openness, you can learn to enhance your awareness and clarity of mind. This can help you to develop greater mindfulness and insight into the nature of reality.

5. ENCOURAGE NON-DUAL AWARENESS

The goal of practice is to cultivate a non-dual awareness that transcends ordinary concepts of self and other. By meditating on the sky, you can learn how to let go of dualistic (black/white) thinking and connect with the inherent unity of all things.

A Final Word ...

By using the Sky Wisdom Oracle deck and meditating with the sky, you can directly experience your true nature and awaken to an enlightened perspective of yourself and the world.

The way of sky wisdom is first to recognize your sky nature, then through practice gain confidence to stabilize and rest within it, and finally to integrate this ultimate dimension of reality into every moment.

I hope you find as much bliss in bringing the sky into your daily life as I do.

Freedom

"Enlightenment will be found by seeing the nature of the mind itself as sky-like. Then there is no other peace to attain."

LONGCHENPA

Breath

BREATH

ASPIRATION May you experience the sky in every breath.

REFLECTION The sky is within us, through us and all around us. The sky cannot be divided. Your breathing is simply a breeze through the open spaces of the sky. Through mindful breathing you can connect your soul with the sky and dissolve all division between inner and outer. The breath can purify the feeling of separation from the outside world.

MEDITATION Visualize your entire body as hollow and sky-like. Now breathe in and imagine the air dissolving into your inner sky. Breathe out and follow your breath as it dissolves into the sky above you. Continue following and dissolving the breath within and without. Eventually the inner and outer skies will merge into the one experience of unity.

MANTRA I breathe in clarity and breathe out all that does not serve me as I rest in the pure nature of being.

Creativity

CREATIVITY

ASPIRATION May you be open to infinite possibilities.

REFLECTION A closed mind doesn't have many options, but an open mind is connected with limitless potential. The source of creativity can be found within the luminous expanse of your mind. Inspiration comes naturally when you're relaxed and open. Rely on your innocent mind to see the world in a clear and novel way.

MEDITATION Trust your sky nature. Listen to the inner silence from where all thoughts arise. Allow your mind to settle in its natural state of luminosity and emptiness. From within the expanse of inner silence feel the spontaneous presence of joy and love. This is the true source of creativity.

MANTRA I am a creative force, bringing my unique gifts and talents to the world.

Flight

FLIGHT

ASPIRATION May you let go and fly.

REFLECTION Worries and stress can weigh us down, putting pressure on us so we feel trapped and oppressed. The terrible irony is that we hold on to the very things that are causing us suffering. Remember you can let go, fly and be free within your sky nature. By simply having the courage to release your last thought you can fly high and find the natural peace in every moment.

MEDITATION Breathe in deeply and exhale slowly. Release the tension in your body on the out-breath. Continue to breathe deeply and slowly and settle your mind until you feel lighter, brighter and you are floating without stress within your sky nature. When you can let go of all tension in your body and mind, clarity will naturally shine through.

MANTRA Letting go of all attachment, I soar up into the boundless freedom of my own true nature.

Flow

FLOW

ASPIRATION May you effortlessly flow by on a gentle breeze.

REFLECTION When the sky is your inner foundation, it doesn't matter where you go because you're always at home. This card asks you to connect with the flow, to ride effortlessly through both the bad times and the good, always feeling connected to something bigger. Being in the flow is being guided by a sense of ease and spaciousness.

MEDITATION Give yourself permission to let go of ambition and drift aimlessly. Practise paying attention to the present moment without any agendas or goals; instead, recognize the pure joy of just being. When sitting in meditation, sit without any aims or expectations. Being in the flow means non-resistance and non-attachment to the present situation.

MANTRA I am flow, weightless and unbound, carried effortlessly by the infinite expanse of my own awareness.

Joy

JOY

ASPIRATION May you overflow with joy.

REFLECTION When your home is always with you, you are never lost. Being in touch with the peace and inner light of your sky nature allows you to be confident and carefree in every situation. You flow through life like a fresh breeze, constantly connected to the true source of blissful freedom.

MEDITATION Let go of fear and desire by remembering the natural perfection of your sky nature. With confidence in your true home, simply enjoy every moment. Play within the expanse of possibilities, knowing your true nature can never be harmed. Be the sky dancing with joy.

MANTRA I am carefree and enjoy each moment without attachment or worry.

Liberation

LIBERATION

ASPIRATION May you be naturally free.

REFLECTION Clouds and weather pass through the sky continuously. There is nothing the sky must do to remove them because they remove themselves. They pass by naturally, just as thoughts and emotions pass when you don't interfere or grasp at them. By abiding in your sky nature, you become effortlessly free. This is called natural liberation.

MEDITATION When you're comfortable dwelling in the sky nature of mind, there's nothing else you need to do. Thoughts and emotions naturally pass and free themselves. Simply rest and experience natural liberation. By doing nothing, you allow your thinking to unravel itself and reveal a natural clarity.

MANTRA I am liberated from all suffering and attain ultimate peace.

Natural Perfection

NATURAL PERFECTION

ASPIRATION May you see things perfectly, as they are.

REFLECTION Just as a clear sky is innately brilliant, the very nature of your mind is naturally perfect and clear. As the essence of your mind is naturally perfect, there's nothing you have to do to create stillness, peace or spaciousness. Simply recognize what's already there and rest within it.

MEDITATION With deep confidence in your perfect sky-like nature of mind, just surrender all striving for or against anything and rest back in your pure sky nature. Simply relax without any contrivance or fabrication, without any effort or resistance. Utterly simple, totally brilliant.

MANTRA My true nature is naturally perfect.

Release

RELEASE

ASPIRATION May you be free from the past.

REFLECTION The action of letting go is like taking a leap of faith into the void, into the unknown. This card is asking you to have the courage to release the past and be in the moment with a fresh perspective. Letting go is the key to being fully awakened.

MEDITATION Continuously drop the last thought and begin again with freshness of mind focused on the present moment. Again and again, release and come home to the now. Letting go is a continuous process throughout the day, it's not a one-off thing. Practise for short sessions many times a day.

MANTRA I release all attachments and find freedom in the moment.

Rest

REST

ASPIRATION May you relax within your sky nature.

REFLECTION We constantly try to find peace and happiness by controlling our life, chasing our tail in a never-ending pursuit of fulfilment. Remember that your sky nature is already perfectly pure and fulfilled. Trust that as you rest, everything is already done and peace is already there. There's nothing to do but melt into your true nature.

MEDITATION Give yourself permission to relax. Then rest and gaze within to the clear and open spaces of your mind. Simply relax within that field of openness and experience contentment. Do nothing else. You are already the pure clarity of joy.

MANTRA I rest in the natural state of my mind and find inner peace and contentment.

Stillness

STILLNESS

ASPIRATION May you experience inner peace.

REFLECTION When your attention is caught up in the ongoing stress of the world, you can easily forget the space and stillness you already possess within. Give yourself the daily gift of remembering your sky nature – that inner space of light and peace. Pause many times throughout the day and connect with the stillness at the core of your being.

MEDITATION Look within and notice the difference between "objects of mind" such as thoughts, emotions or sensations, and the mind itself that is aware of those things. Realize how the nature of thoughts, emotions and so on is to come and go, and that awareness itself is free from coming and going because it's always present. Notice the stillness and rest within it.

MANTRA Resting in the silent expanse of my own awareness, I experience stillness in the boundless spaciousness of my own true nature.

Surrender

SURRENDER

ASPIRATION May you give up struggling – and relax.

REFLECTION Life can be filled with ongoing battles and never-ending struggles. It's important to pause regularly, offer yourself a cease-fire and surrender into full acceptance of the present moment. We cannot attain peace through force. It's only by surrendering our resistance to what is that we can find peace.

MEDITATION Look up into the sky and smile. Remember your own perfectly pure nature is always within you even on difficult days. Knowing that your essential pure nature is already free and blissful, give up all striving and simply be at peace in the moment. Rest in the freedom of just being. Surrender to your true nature.

MANTRA I surrender to the flow of life and trust in the clarity and spaciousness of sky wisdom.

Transcendence

TRANSCENDENCE

ASPIRATION May you soar like an eagle.

REFLECTION An eagle soaring above a mountain is at home floating in the sky, free from the concerns far below. In the same way, when you're at peace being in the sky-like nature of mind, you're free from the fluctuations of the world, drawing upon the higher perspective of an eagle soaring. This is freedom through transcendence.

MEDITATION Lift your attention to the sky, into the spaces above you. Spend a moment exploring the completely empty space above and merge your mind with that experience of spaciousness. When you're ready, breathe in that space and totally integrate it into the present moment. Feel the lightness of soaring free, high in the sky.

MANTRA I am free to be my authentic self and live in harmony with all beings.

Clarity

"The blue sky is a seamless combination of light and space, the same way your enlightened mind is the combination of awareness and emptiness."

LONGCHENPA

Awakening

AWAKENING

ASPIRATION May you have sky wisdom in every moment.

REFLECTION The spacious, open and free sky dimension is our fundamental source and deepest identity. When you recognize this deepest dimension of reality is connected with everything, you can become whole and awaken to your sky wisdom. Awakening is knowing who you really are, including your inner being that is spacious, vast and eternal.

MEDITATION Look up to the sky and merge your mind with the open expanse of space and light. Allow your thoughts to dissolve into this vastness of the sky. Without thoughts, notice the lack of division between you and the world. Awaken from the dream of thoughts into the clarity of oneness.

MANTRA I am awake to the truth of my being and the interconnectedness of all things.

Bliss

BLISS

ASPIRATION May you be spontaneously joyful.

REFLECTION Clinging and grasping on to an ego identity causes separation and suffering. The direct antidote to that suffering is to connect with your egoless sky nature. Being open and egoless is joyful. Bliss is the natural state of your sky nature when the ego is absent.

MEDITATION Put down your distractions, detach from worldly pleasures and relax into the natural bliss of just being. Start with a deep sigh, saying "Ahhh". Then practise the "three simples" of natural bliss meditation: simply sitting, simply breathing and simply being. Experiencing bliss can truly be that simple.

MANTRA I experience the boundless bliss of my own true nature, overflowing with joy and love, at one with all of existence.

Brightness

BRIGHTNESS

ASPIRATION May you have the brightness of a clear blue sky.

REFLECTION When you can rise above a situation, you can connect with sky wisdom and see the bigger picture, the broader perspective. But it's not your eyes that see, it's your mind. Clarity is not what you're looking at, it's where you're looking from. Clear your mind and be like the bright, cloudless sky.

MEDITATION True seeing is knowing your own nature. Look within your mind for the gaps between your thoughts. Notice the indefinable open nature of the spaces between thoughts. See for yourself that your fundamental nature is empty and clear. Thoughts come and go but the sky-like awareness is always present.

MANTRA I shine, illuminating all that is and embodying the brightness of my enlightened mind.

Brilliance

BRILLIANCE

ASPIRATION May you be the radiance of the world.

REFLECTION The brilliant light in you is invisible until you shine it upon the world, just as the sunlight streaming through a blue sky is invisible until it meets the Earth and illuminates all things. By connecting with the stillness and openness of your inner sky mind, you also connect with the presence of brightness, with the brilliant radiance that you are. Wherever you go there is light.

MEDITATION Close your eyes and direct your attention inward. Notice the non-moving mind – that still point at the core of your awareness. Keep concentrating until you connect with the inner space of openness that never changes. Once you've found the stillness, recognize that there is always a presence with it. The emptiness is always connected to awareness. The stillness always has brilliance.

MANTRA May the brilliant light of my inner wisdom shine forth and guide me and all beings toward awakening.

Enlightenment

ENLIGHTENMENT

ASPIRATION May sky wisdom always be with you.

REFLECTION We all have an enlightened nature of bliss, love and wisdom. But we don't see it or experience it because we are distracted by the stress, drama and desires of the ego. If we really do have an enlightened nature, shouldn't we dedicate at least some time to finding it? Pause throughout the day, put distractions aside and focus on that space within that has no boundaries.

MEDITATION Know yourself as sky nature instead of a fixed and separate ego. This open nature is at the heart of all experiences. Even during the roughest weather, the pure sky is still present. At all times and through all emotional and mental states, continually remember the refuge of your sky nature within and come from that place of spaciousness and clarity.

MANTRA In the boundless expanse of my own awareness, I awaken to the true nature of reality beyond concepts.

Focus

ASPIRATION May you never be distracted from sky wisdom.

REFLECTION What you pay attention to becomes your world. The more you focus on negative things, the more stress and anxiety you'll have. The more you pay attention to your sky nature, the more fresh, calm and directed your mind will be. So focus on your open, clear and spacious nature.

FOCUS

MEDITATION Try to spend many small moments refining your attention during the day by resting in the naturalness of your sky nature. With effortless ease allow your focus to relax and expand into the present moment. With nothing left to do, just be and your mind will be naturally focused.

MANTRA I focus my attention on what is truly important and let go of negativity.

Intuition

INTUITION

ASPIRATION May you be guided by sky wisdom.

REFLECTION In the expanse of light within, there are all the answers in the universe. The quieter your mind, the more you can hear the whispers of sky wisdom. Intuition comes from the peaceful stillness of your sky nature. The quieter you are, the more you can hear.

MEDITATION Before making any decisions, from small ordinary choices to big life-changing ones, always stop, breathe and connect with the stillness of the sky. Allow your intuitive wisdom to emerge naturally from this stillness, and trust its guidance. The inner space is not just a void, it's full of wisdom and joy.

MANTRA I trust my intuition and inner wisdom to guide me on my path.

Light

LIGHT

ASPIRATION May the light of awareness always shine bright.

REFLECTION Just as the blue sky might appear empty when it is suffused with light, your own sky nature is empty but filled with awareness. Sky nature is the union of the light of awareness and the open spacious field of your mind. Light is a symbol for your awareness.

MEDITATION Turn your attention around onto itself. Notice your own light of awareness that is always shining, always knowing. Throughout the day look up at the sky and remember that awareness is at the core of all experiences and has the same open spacious nature as the sky. Merge your awareness with the sky.

MANTRA In the radiant clarity of my own awareness, I embody the light of wisdom and compassion.

Lucidity

LUCIDITY

ASPIRATION May you attain natural, fresh wakefulness.

REFLECTION Lost in clouds of thoughts the mind forgets the simplicity and brilliance of clear awareness. Just like being lost in a dream, you forget it's all your own mind. Awakening from the dream of thought brings you to the luminous clarity of natural awareness.

MEDITATION By clearly being aware in the present moment without judging or clinging to it, rest in the clarity of what is actually happening rather than your thoughts about it. The clarity of the sky is within all things. By simply being lucid and mindful, you allow your sky wisdom to radiate naturally.

MANTRA The innate lucidity of my mind shines eternally.

Mindfulness

MINDFULNESS

ASPIRATION May you always live in the present moment.

REFLECTION It's all too easy to live in the past through regret and nostalgia, or live in the future through hope and fear. This card is asking you to bring your attention fully to the present moment and enjoy life right now.

MEDITATION Throughout the day, practise mindful breaths by drawing your attention away from the thinking mind and focusing your attention on the next breath.

BREATHE in the sky and bring your mind home to this present moment; exhale the sky and let go and relax. Allow your in-breath to draw all your attention to the now, then let your out-breath relax all tension and just be present.

MANTRA In every moment I am clearly present to all things, while allowing them to be as they are.

Transparency

TRANSPARENCY

ASPIRATION May you see through all illusions.

REFLECTION When you become familiar with paying attention to your sky nature, your mind is naturally expansive and calm. This empowers your sky wisdom to clearly see deceptions and illusions that try to ensnare your attention in narrow and limited perspectives.

MEDITATION Continually cut through the fixations and ruminations of your mind to arrive back home in the open transparent spaces of your pure nature, your natural state. You don't have to create this naturally free spaciousness; you just have to break through the barriers you've built up around it and rest as your authentic self.

MANTRA I am transparent and authentic, expressing my true self without fear or judgement.

Untouched

UNTOUCHED

ASPIRATION May your mind be like a pristine blue sky.

REFLECTION Your soul's deepest nature is clear and untouched like the sky. This sky nature is also sharply aware. It's the openness of mind that naturally allows the pristine clarity to shine forth. Emptiness and clarity go together as one, just as the space and light of the sky co-exist.

MEDITATION Let go of all effort and simply abide in the natural state of mind, and its pristine clarity will shine forth. As you rest in your sky nature things become clearer, your perspective broadens and wisdom naturally reveals itself. Simply be at ease – the clouds will blow over and clarity will come naturally.

MANTRA My mind is untouched and clear, reflecting my true nature.

Spaciousness

*"Mind itself is a vast expanse,
the realm of unchanging space."*

LONGCHENPA

Absence

ABSENCE

ASPIRATION May you fully enjoy nothingness.

REFLECTION The absence of stress is peace; the absence of noise is silence; and the absence of movement is stillness. Usually, our attention is totally caught up in stress, noise and movement but the sky reminds us of our natural potential for peace, silence and stillness within. In the absence of thoughts, peace is naturally there.

MEDITATION Connect with the sky and notice its qualities of deep stillness and spaciousness. Use that as a portal to notice your own inherent stillness and inner sky. Simply rest in that natural state of emptiness and you'll find it's full of joy and happiness.

MANTRA I find peace in the absence of all that is not truly essential.

Awareness

AWARENESS

ASPIRATION May you be the witness.

REFLECTION Mystics throughout the ages have reminded us that awareness itself is our fundamental identity. Your body is impermanent, your thoughts are temporary and your emotions are fleeting. It is awareness that is the constant thread behind all these things.

MEDITATION Become aware of awareness. Notice that constant quality of knowing, the quality of being the witness. See for yourself that whenever you look, it's always there. However, when you investigate deeply, you realize this awareness cannot be pinned down as it is open like the sky. Explore again and again this indefinable sky-like nature of your own awareness until you know it to be your fundamental identity.

MANTRA I am awareness itself – open, clear and the pure witness of whatever appears.

Expansion

EXPANSION

ASPIRATION May you expand your consciousness.

REFLECTION At times your mind and soul can feel constricted and trapped. The opposite is to connect your inner being with the sky and feel the experience of expansion. By raising your consciousness into the sky, you can reconnect with a joyful and open state of mind.

MEDITATION To experience being expansive you simply have to let go and relax – your mind will naturally feel freer and more open. Drop the heavy thoughts weighing you down. Breathe deeply while looking up into the sky and set your soul free. Expansion is your natural state.

MANTRA I expand my consciousness and embrace the interconnectedness of all beings.

Infinity

INFINITY

ASPIRATION May you see beyond all limits.

REFLECTION Your sky nature is your connection with eternity. It has no limits or boundaries; it is completely open and has no fixed position. The ego, on the other hand, is fixed and definable and feels like it's trapped inside your body. To connect with the infinite is to go beyond the ego and fly in the open expanse of the sky.

MEDITATION Thoughts are always limitations. To get a glimpse of thought-free awareness, take a deep breath in, breathing in the sky. Briefly hold the breath and look deeply into the vast sky within. By momentarily stopping the breath you can stop thoughts and gain a glimpse of an infinite sky-like experience.

MANTRA I am infinite and boundless, connected to the vast universe.

Love

LOVE

ASPIRATION May your heart fill the sky.

REFLECTION Love is space. Just as the sky embraces everything equally and is the space for all experiences to unfold, divine love holds space for the existence of all beings. As infants our love is limited to the ones around us, but as we mature it's possible for our love to include all beings and we can embrace all situations.

MEDITATION Hold your hand over your heart, breathe deeply and offer your full love and acceptance to yourself. Imagine your heart glowing with divine light as it starts to expand and grow. Send your light into the infinite sky with the intention that it blesses everything and everyone. Merge your love with the sky. May all beings be free; may all beings be happy.

MANTRA My love embraces and accommodates all beings under the sky.

Non-judgement

NON-JUDGEMENT

ASPIRATION May your mind be free of judgements.

REFLECTION This card is inviting you to drop judgements and expectations and instead find that space within you that is happy and comfortable to simply be, let be and let live. This is the sky wisdom of an open, spacious mind that doesn't cling to the ebb and flow of opinion.

MEDITATION Allow your judgements and opinions to be like clouds floating by – just temporary, fleeting and unimportant. Sit still for a moment and connect with the open expanse of the sky with confidence; you're naturally perfect already without having to prove or defend anything.

MANTRA I am non-judgemental and accept all beings as they are.

Openness

OPENNESS

ASPIRATION May you have an open mind.

REFLECTION Sky wisdom calls us to fulfil our potential to hold space, to embody a field of open acceptance that is receptive and responsive. Do this by adopting an open stance of listening and a posture of complete innocence and non-judgement. Openness is a primary enlightened quality of sky wisdom.

MEDITATION Actively let go of opinions throughout the day. Remember your perspective is relative to your situation and allow space for others to hold differing views. After cutting through your opinions and perspectives, rest in the spaces that open up, the authentic nature beyond thoughts of right or wrong.

MANTRA I am open to new experiences and perspectives, embracing the richness of life.

Potential

POTENTIAL

ASPIRATION May you be free from limiting beliefs.

REFLECTION Fulfilling your potential for joy, love and wisdom in every moment is as natural as the grass growing. But limiting beliefs, doubts and opinions can block your light from shining fully. To achieve your potential, stop listening to negative thoughts and instead look to the vast spaces of the inner sky for inspiration.

MEDITATION Notice when negative thinking takes hold of your mood and your mind, dragging you into darkness and limitations. As soon as this happens, lift your attention to the sky and focus on the space and light. This will clear your mind and reconnect you with your natural potential, your natural freedom.

MANTRA I am aware of my full potential and live with purpose and intention.

Unbound

UNBOUND

ASPIRATION May your mind be limitless.

REFLECTION The very ground of your being right now has no boundaries, it's limitless. This can be scary for the ego, but it is utterly liberating for the soul. This card is asking you to skydive into the void and let go of all fixed identities.

MEDITATION Drop your last thought and simply relax back in your sky nature of undivided expansiveness. There's nothing to do, you're already unbound. Unwind and completely let go, so that you ease into your natural and authentic state of being.

MANTRA My true nature is unbound and infinite, beyond all limitations.

Unity

UNITY

ASPIRATION May you be one with all people under the sky.

REFLECTION The sky unites us all. It's above the head of everyone living on the planet. And it's the inner nature of the mind and soul of all people. Beyond gender, race and religion there is a common nature of all conscious beings – sky nature. The vast intangible essence of our own mind is the same in all beings. We are all one.

MEDITATION Remember to see the sky in all beings. As you rest in the peace and tranquillity of your own sky nature, remember that it's the ultimate identity of all people. On the deepest level we are all the same, we are all united. Our true source is essentially the same nature.

MANTRA I recognize the unity of all things and live in harmony with all people.

Unknowing

ASPIRATION May you discover the wisdom of not knowing.

REFLECTION To think "outside the box" in unique and creative ways requires you to detach from the norms and pre-set patterns of knowledge. Not knowing something or having an open beginner's mind can help you see things others can't. It's the not knowing that creates the open spaces for true wisdom to dawn. The clear mind understands the deepest concepts.

MEDITATION Develop confidence in resting in an open sky mind – an easy-going way that isn't compelled to cast judgements or express opinions. Just be happy being simple, open and free in the moment, not knowing anything. Make this your home; rest in sky wisdom as your default state of being.

MANTRA I embrace the mystery of life and find freedom in the unknown.

Wholeness

WHOLENESS

ASPIRATION May you be the totality of the moment.

REFLECTION The sky element unifies the world. Everything occupies space or the sky. Space surrounds everything and runs through everything. Space makes up 99.9 per cent of all atoms' volume and everything is made of atoms. When you understand that the space in the sky connects everything, you can begin to experience the wholeness of everything and everyone as part of the same unified field.

MEDITATION Look up at the sky or even just think about and visualize it. Connect with the enormous, unified field that the sky is. Stretch your imagination and allow yourself to be the sky without a centre. You are everywhere all at once, at the heart of all beings and the core of all things. Practise identifying yourself with this field without centre rather than something fixed and localized.

MANTRA I am complete and whole, and I honour the interconnectedness of all beings in universal oneness.

Essence

"The nature of the mind is the ultimate sphere like a boundless sky; the nature of the sky is the nature of the mind, meaning they are not separate, they are the oneness of enlightened presence."

LONGCHENPA

Authenticity

AUTHENTICITY

ASPIRATION May you express yourself freely and honestly.

REFLECTION Being in alignment with your sky nature allows the free flow of expression. When the very core of your being is the empty, clear and bright space of the sky, you are able to be your authentic self. The key is identifying with the clarity of openness that is indestructible. Since the open space of your essential nature cannot be harmed, you can be yourself completely without fear.

MEDITATION Breathe deeply and slowly, connecting with the permanent space of your inner sky nature. Simply by breathing and just being you can align with the eternal presence of light and space, which is the essential nature of your mind right now. Practise just being, which is pure authenticity. You have nothing to prove and nothing to defend; just breathe and simply be.

MANTRA I am grounded in my authentic being, radiating love and light as I contribute to the beauty and harmony of the world around me.

Discovery

DISCOVERY

ASPIRATION May you discover your sky nature.

REFLECTION Time can easily go by while you are lost in thoughts – entangled in worries and stress, you lose sight of the natural freedom you already possess. But just like turning on a light can immediately dispel darkness, discovering and recognizing your sky nature immediately cuts through these entanglements.

MEDITATION Pause for a moment. Breathe long slow breaths. Imagine your in-breath is like a strong wind blowing away the inner clouds of stress, doubt and worry. After a few moments the clouds have gone, completely revealing the pure, clear sky nature within. Know it's your true home.

MANTRA I awaken from the dream of thoughts and discover my eternal, radiant, true nature.

Nature

NATURE

ASPIRATION May you experience your true self.

REFLECTION Your essential sky nature can never be harmed or improved; it is perfect just the way it is. This is your innate goodness. Your fundamental nature is pure, open and unconfined like the sky. By developing confidence in this fact, you will loosen the grip of stress and anxiety over you so that you can fly free.

MEDITATION With total confidence in your pure sky nature, give up all effort and struggle and be at ease by simply relaxing back into this union of space and light. You can always rely on your sky nature because it's always there. Just rest back, let go and surrender to its innate purity, which is the very essence of your being – your true self.

MANTRA My true essence is pure and radiant, reflecting the ultimate truth of reality.

Omnipotent

OMNIPOTENT

ASPIRATION May the sky always be present within you.

REFLECTION Wherever you go the sky is always above you. Similarly, your mind's true nature is always present within you. By consistently recognizing and connecting with the inner spaces of your mind's natural state, you will lessen the harmful effects of ego and instead integrate sky wisdom into everything you do.

MEDITATION Remember your sky nature. For short periods many times a day, recall your true nature as openness and light, emptiness and awareness. By being mindful of this expansive inner experience as much as possible, you will transform your every action and enlighten your heart and soul wherever you go. Your sky nature is always present and always perfect; you just have to notice it and integrate it into your life.

MANTRA I am powerful and capable of achieving my highest potential in every moment.

Peace

PEACE

ASPIRATION May the serenity of the sky be with you.

REFLECTION The sky never moves, it is always still. In the same way, your sky nature is always still and therefore peaceful. You don't have to make it peaceful because it already is. You just have to connect and rest within that already open, still and peaceful sky space within. This is your natural peaceful state of being.

MEDITATION Any effort or trying to be peaceful is ultimately unsuccessful. Any resistance to what is happening creates subtle tension. Therefore, total surrender and acceptance of the present moment is key to finding peace. Stop trying, stop controlling and stop resisting; simply rest in your true sky nature of great peace.

MANTRA In the boundless spaciousness of my own true nature, I am at peace, resting in the ease of the present moment.

Perspective

PERSPECTIVE

ASPIRATION May you always include the higher perspective.

REFLECTION Beyond thoughts of right and wrong there is an open field of just being; this is your true home. Just as the sky is always there, it's always possible to rise above any situation and come from that place of expansive awareness and wider perspective that transcends all limitations.

MEDITATION Imagine floating up into the sky further and further away until your stress and problems look small and insignificant below. Don't ignore your issues, just don't bring them to the fore, so you can gain some distance and perspective. Your sky wisdom will naturally emerge and you'll find space and a way through.

MANTRA I transcend all ideas and limitations and from this fresh perspective find freedom in my true nature.

Purity

PURITY

ASPIRATION May you know your pure nature.

REFLECTION There's no amount of bad weather that can ever stain the sky. It retains its purity even amid the harshest of conditions. Similarly, your inner sky nature is always pure from the beginning and can never be harmed by temporary circumstances. This is your indestructible pure nature.

MEDITATION On a clear day go to a high place with a good view of the sky or simply look up. For a few moments breathe deeply and stare into the blue clarity of the sky. Merge your mind into the vast purity of the sky until you recognize or feel that it's also your nature.

MANTRA Within the luminous expanse of my own awareness, I rest in the pure, natural sky nature, free from all confusion and delusion.

Self-awareness

SELF-AWARENESS

ASPIRATION May you know yourself fully.

REFLECTION Self-awareness is usually limited to what you can be aware of about yourself – your thoughts, reactions, emotions or bodily sensations. However, the deepest awareness is of awareness itself. You are really looking for what is doing the looking. Self-awareness is ultimately knowing your sky nature as your true self.

MEDITATION Pay attention to your inner world. Notice your thoughts and how you feel; note the sensations in your body and also your mood. Now make the simple enquiry: what is it that's noticing these things? And shift your attention to become aware of awareness itself. Know yourself to be that awareness.

MANTRA I am fully present in the moment, attuned to my deepest nature – the luminous expanse of my own awareness.

Sky

SKY

ASPIRATION May you be the sky and accept the weather.

REFLECTION Accept thoughts and strong emotions as temporary weather patterns that move through your sky nature. You can begin to distinguish between temporary mental events and the permanent sky-like awareness that never changes. You are that permanent awareness.

MEDITATION Identify with openness and allow everything to be "as it is", while knowing thoughts and emotions will pass through by themselves. There's absolutely nothing to do and nothing to change; simply rest without effort in your sky mind and enjoy the passing conditions with confidence in your ultimate nature.

MANTRA My true nature is like the sky – vast, open and free.

Trust

TRUST

ASPIRATION May you be free from the clouds of doubt.

REFLECTION Once you've had a glimpse of your nature as open, pure and clear like a blue sky, you can learn to trust and rest in this wisdom. You can rely on an open mind rather than opinions, and trust that "just being" is naturally perfect. This is different from having faith because through meditation you can experience and know it for yourself.

MEDITATION When lying on the floor you can trust the ground to support you. Similarly, you can let go of the striving and struggling of the ego and rely on your own sky wisdom to support you. Sit still and relax your mind, surrender to the clear and open sky nature of your being and trust the wisdom that emerges.

MANTRA I trust in the fundamental goodness and wisdom of the universe as I rest in the natural state of my own mind.

Vastness

VASTNESS

ASPIRATION May you find humility in facing the sky.

REFLECTION The ego always thinks it's right and also the biggest and best thing in the room. By connecting with something larger than the ego you can start to loosen its suffocating grip on you. You can do this by simply connecting with the vastness of the sky and putting your ego into perspective.

MEDITATION Breathe in deeply and imagine your mind and the sky mingle and unite within. On the out-breath totally relax and imagine or feel your mind expanding beyond the body and out into the vastness of the sky. Continue to sky-breathe until you feel the shell of the ego crack open and your soul expand into the vast space.

MANTRA My awareness is vast and all-encompassing, embracing all beings and phenomena.